One-Dimensional Transport with Equilibrium Chemistry (OTEQ): A Reactive Transport Model for Streams and Rivers

By Robert L. Runkel

Toxic Substances Hydrology Program

Techniques and Methods Book 6, Chapter B6

U.S. Department of the Interior
U.S. Geological Survey

U.S. Department of the Interior
KEN SALAZAR, Secretary

U.S. Geological Survey
Marcia K. McNutt, Director

U.S. Geological Survey, Reston, Virginia: 2010

For product and ordering information:
World Wide Web: http://www.usgs.gov/pubprod
Telephone: 1–888–ASK–USGS

For more information on the USGS—the Federal source for science about the Earth,
its natural and living resources, natural hazards, and the environment:
World Wide Web: http://www.usgs.gov
Telephone: 1–888–ASK–USGS

Suggested citation:
Runkel, R.L., 2010, One-dimensional transport with equilibrium chemistry (OTEQ) — A reactive transport
model for streams and rivers: U.S. Geological Survey Techniques and Methods Book 6, Chapter B6, 101 p.

Contents

Abstract 1

1 Introduction 2
   1.1 Overview 2
   1.2 Applicability 2
   1.3 Related Reading 2
   1.4 Report Organization 2
   1.5 Acknowledgments 3

2 Theory 4
   2.1 Overview 4
   2.2 Conceptual Model, Governing Equations, and the Sequential Iteration Method 5
      2.2.1 Model Assumptions 5
      2.2.2 Derivation of Governing Equations 6
      2.2.3 A General Solution Scheme based on Sequential Iteration 8
   2.3 Process Formulation 11
      2.3.1 pH 11
      2.3.2 Precipitation/Dissolution 11
      2.3.3 Sorption 13
      2.3.4 Oxidation/Reduction 20
      2.3.5 Transient Storage 22
      2.3.6 Settling of Solid Phases 24
   2.4 Numerical Solution 24
      2.4.1 The Conceptual System — Segmentation 24
      2.4.2 Boundary Conditions 25
      2.4.3 Initial Conditions 26

3 User's Guide 27
   3.1 Conceptual System, Revisited 27
   3.2 Input/Output Structure 28
   3.3 Input Format 29
      3.3.1 Units 30
      3.3.2 Internal Comments 30
      3.3.3 The Control File 30
      3.3.4 The Parameter File 31
      3.3.5 The Flow File 40
      3.3.6 The MINTEQ Input File 43
      3.3.7 The MINTEQ Database Files 45
   3.4 Input File Preparation and Model Execution 45
      3.4.1 Preparation of the Control File 45
      3.4.2 Preparation of the Parameter and Flow Files — Use of MINTEQ 45
      3.4.3 Preparation of the MINTEQ input file — Use of PROTEQ 46
      3.4.4 Execution of OTEQ 47

3.5   Output Analysis 48
      3.5.1   The Solute and Solid Output Files 48
      3.5.2   Concentration-Distance Output Files 49
      3.5.3   The Post-Processor, POSTEQ 49
      3.5.4   Plotting Alternatives 49
4   Model Applications 50
   4.1   Application 1: Time-Variable Simulation of a Solute Pulse with Precipitation 50
      4.1.1   The Control File — Application 1 51
      4.1.2   The Parameter File — Application 1 51
      4.1.3   The Steady Flow File — Application 1 54
      4.1.4   The MINTEQ Input File — Application 1 54
      4.1.5   Simulation Results — Application 1 55
      4.1.6   Numerical Issues — Application 1 56
   4.2   Application 2: Time-Variable Simulation of pH and pH-Dependent Precipitation . . . 57
      4.2.1   The Parameter and Flow Files — Application 2 58
      4.2.2   The MINTEQ Input File — Application 2 59
      4.2.3   Simulation Results — Application 2 59
   4.3   Application 3: Time-Variable Simulation of Copper Sorption to the Streambed 61
      4.3.1   The Parameter File — Application 3 61
      4.3.2   The Unsteady Flow File — Application 3 64
      4.3.3   The MINTEQ Input File — Application 3 64
      4.3.4   Simulation Results — Application 3 66
      4.3.5   Numerical Issues — Application 3 66
   4.4   Application 4: Steady-State Simulation of Existing Conditions and Remedial Action 68
      4.4.1   Quasi-Steady-State Simulations 68
      4.4.2   Modeling Existing Conditions and Remediation 69
      4.4.3   Simulation Results — Application 4 70
   4.5   Application 5: Steady-State Simulation of Sorption onto Water-Borne Precipitates 70
      4.5.1   The Parameter, Flow, and MINTEQ Input Files — Application 5 72
      4.5.2   Specification of H and CO_3: The Case of Waterborne Solid Phases 73
      4.5.3   Simulation Results — Application 5 74
5   Software Guide 75
   5.1   Supported Platforms 75
   5.2   Software Distribution 75
   5.3   Installation 76
      5.3.1   Creating the OTEQ Directory Structure 76
      5.3.2   Updating the User's Path 76
      5.3.3   Creating User Work Areas 77
   5.4   Compilation 78
   5.5   Software Overview 78
      5.5.1   Model Development 78
      5.5.2   Include Files 79
      5.5.3   Error Checking 80
References Cited 81
Glossary 83
Appendix 1. Modifications to the MINTEQ Database 86

Figures

1 Conceptual surface-water system used to develop the governing differential equations.... 6
2 The sequential iteration approach for the reactive surface-water model. 9
3 Computation of the dissolution source/sink term. 12
4 Conceptual surface-water system for sorption to a static surface. 15
5 Use of the equilibrium submodel for a static surface. 16
6 Conceptual surface-water system for sorption to a dynamic surface.. 17
7 Conceptual surface-water system for sorption to static and dynamic surfaces........... 18
8 Use of the equilibrium submodel for static and dynamic surfaces.. 19
9 Iterative scheme for oxidation/reduction. 21
10 Conceptual surface-water system used to develop the governing differential equations. 23
11 Segmentation scheme used to implement the numerical solution.. 24
12 Upstream boundary condition defined in terms of a fixed concentration. 25
13 Downstream boundary condition defined in terms of a fixed dispersive flux............. 25
14 Conceptual system that includes one or more reaches............................. 27
15 The first reach in the conceptual system and the required input variables. 28
16 OTEQ Input/Output files. 29
17 Upstream boundary condition options.. 39
18 Example solute output file. 48
19 Example precipitate output file, for the case of PRTOPT=1........................... 49
20 Upstream boundary condition for double pulse injection. 50
21 Control file for Application 1. 51
22 Parameter file for Application 1, record types 1–15. 52
23 Parameter file for Application 1, record types 16–28. 53
24 Steady flow file for Application 1. 54
25 MINTEQ input file for Application 1. 55
26 Simulated concentrations of calcium (Ca) and sulfate (SO$_4$) at 200 meters. 56
27 Peak total waterborne sulfate concentration at 200 meters. 57
28 Partial listing of the parameter file for Application 2................................. 58
29 Simulated and observed concentrations of (a) total dissolved iron and (b) dissolved aluminum. 60
30 Segmentation scheme for the hypothetical stream with tributary input 62
31 Partial listing of the parameter file for Application 3.. 63
32 Unsteady flow file for Application 3. 65
33 Copper concentrations resulting from copper sulfate treatment...................... 66
34 Spatial profiles of simulated copper concentration at 4 hours........................ 67
35 Time required to reach quasi-steady-state conditions.. 69
36 Spatial profiles: pH, iron, and aluminum (existing conditions and remediation).......... 71
37 Partial listing of the parameter file for Application 5.. 72
38 Spatial profiles of simulated and observed: pH, iron, and arsenic..................... 74
39 OTEQ directory structure... 77

Tables

1　The OTEQ control file. 30

2　The parameter file — record types 1–10. 32

3　The parameter file — record type 11. 33

4　The parameter file — record type 12. 33

5　The parameter file — record type 13. 34

6　The parameter file — record type 14. 34

7　The parameter file — record type 15. 34

8　The parameter file — record type 16. 35

9　The parameter file — record types 17–18. 35

10　The parameter file — record types 19–20. 36

11　The parameter file — record type 21. 36

12　The parameter file — record type 22. 36

13　The parameter file — record types 23 and 24. 37

14　The parameter file — record types 25–26. 38

15　The parameter file — record type 27. 38

16　The parameter file — record type 28, upstream boundary conditions. 39

17　Steady flow file — record type 1. 40

18　Steady flow file — record type 2. 40

19　Steady flow file — record type 3. 41

20　Unsteady flow file — record type 1. 41

21　Unsteady flow file — record types 2 and 3. 42

22　Unsteady flow file — record types 4–7. 42

23　The MINTEQ input file — record types 1–6. 44

24　The MINTEQ input file — record types 7 and 8. 44

25　The MINTEQ input file — record type 9. 44

26　Sorption parameters — hydrous ferric oxide (HFO; Dzombak and Morel, 1990). 47

27　Stand-alone MINTEQ runs to reproduce observed pH of 4.43. 73

28　Supported systems. 75

29　Files to download. 75

30　Development environments. 78

31　Maximum dimensions and default values from fmodules.inc. 79

32　Aqueous species for which MINTEQ version 3 enthalpy and logK values are identical to those in wateq4f.dat. 87

33　Aqueous species for which MINTEQ version 3 enthalpy and logK values differed from those in wateq4f.dat. 92

34　Mineral species for which MINTEQ version 3 enthalpy and logK values are identical to those in wateq4f.dat. 93

35　Mineral species for which MINTEQ version 3 enthalpy and logK values differ from those in wateq4f.dat. 100

Conversion Factors

Multiply	By	To obtain
Length		
centimeter (cm)	0.3937	inch (in.)
millimeter (mm)	0.03937	inch (in.)
meter (m)	3.281	foot (ft)
kilometer (km)	0.6214	mile (mi)
Area		
square meter (meter^2)	0.0002471	acre
square kilometer (kilometer^2)	247.1	acre
square centimeter (centimeter^2)	0.001076	square foot (ft^2)
square meter (meter^2)	10.76	square foot (ft^2)
square centimeter (centimeter^2)	0.1550	square inch (in^2)
Volume		
liter	33.82	ounce, fluid (fl. oz)
liter	0.2642	gallon (gal)
cubic meter (meter^3)	264.2	gallon (gal)
cubic centimeter (centimeter^3)	0.06102	cubic inch (in^3)
cubic meter (meter^3)	35.31	cubic foot (ft^3)
cubic meter (meter^3)	0.0008107	acre-foot (acre-ft)
Flow rate		
cubic meter per second (meter^3/s)	70.07	acre-foot per day (acre-ft/d)
meter per second (meter/s)	3.281	foot per second (ft/s)
cubic meter per second (meter^3/s)	35.31	cubic foot per second (ft^3/s)
liter per second (liter/s)	15.85	gallon per minute (gal/min)
cubic meter per day (meter^3/d)	264.2	gallon per day (gal/d)
Mass		
gram (g)	0.03527	ounce, avoirdupois (oz)
kilogram (kg)	2.205	pound avoirdupois (lb)
Energy		
joule (J)	0.0000002	kilowatt hour (kWh)

Temperature in degrees Celsius (°C) may be converted to degrees Fahrenheit (°F) as follows:

°F = (1.8 x °C) + 32

Temperature in degrees Fahrenheit (°F) may be converted to degrees Celsius (°C) as follows:

°C = (°F - 32) / 1.8

Specific conductance is given in microsiemens per centimeter at 25 degrees Celsius (µS/cm at 25°C).

Concentrations of chemical constituents in water are given either in milligrams per liter (mg/liter) or micrograms per liter (µg/liter).

One-Dimensional Transport with Equilibrium Chemistry (OTEQ): A Reactive Transport Model for Streams and Rivers

By Robert L. Runkel

Abstract

OTEQ is a mathematical simulation model used to characterize the fate and transport of waterborne solutes in streams and rivers. The model is formed by coupling a solute transport model with a chemical equilibrium submodel. The solute transport model is based on OTIS, a model that considers the physical processes of advection, dispersion, lateral inflow, and transient storage. The equilibrium submodel is based on MINTEQ, a model that considers the speciation and complexation of aqueous species, acid-base reactions, precipitation/dissolution, and sorption.

Within OTEQ, reactions in the water column may result in the formation of solid phases (precipitates and sorbed species) that are subject to downstream transport and settling processes. Solid phases on the streambed may also interact with the water column through dissolution and sorption/desorption reactions. Consideration of both mobile (waterborne) and immobile (streambed) solid phases requires a unique set of governing differential equations and solution techniques that are developed herein. The partial differential equations describing physical transport and the algebraic equations describing chemical equilibria are coupled using the sequential iteration approach. The model's ability to simulate pH, precipitation/dissolution, and pH-dependent sorption provides a means of evaluating the complex interactions between instream chemistry and hydrologic transport at the field scale.

This report details the development and application of OTEQ. Sections of the report describe model theory, input/output specifications, model applications, and installation instructions. OTEQ may be obtained over the Internet at http://water.usgs.gov/software/OTEQ.

1   Introduction

1.1   Overview

The study of solutes in streams and rivers is inherently complex. A multitude of physical, biological, and geochemical processes influence solute fate and transport. Study of individual processes is confounded by the complex interaction between physical transport processes that act to move solutes downstream and the biogeochemical processes that influence chemical speciation. Individual processes may be studied by employing simulation models that describe process dynamics in a mathematical framework.

Studies of solute fate and transport commonly employ transport models that describe the physical processes of advection and dispersion and some specific chemical and biological reactions (for example, Bencala, 1983; Kuwabara and others, 1984; Brown and Hosseinipour, 1991; Chen and others, 1996). These models describe chemical speciation and sorption using kinetic rate constants and empirical partition coefficients. This general approach is limited in that the database of kinetic rate constants is strikingly sparse. In addition, many sorption reactions are thought to adhere to more mechanistic sorption models (for example, surface complexation). Although these transport models provide an accurate description of physical transport, they often do not include the degree of chemical sophistication needed to describe pH-dependent processes. Chemical equilibrium models, meanwhile, describe pH-dependent reactions in batch systems, but do not consider transport. Fortunately, many chemical reactions are sufficiently fast so that local equilibrium may be reasonably assumed. It is therefore possible to develop a coupled model wherein a transport model is used to describe physical processes and a chemical equilibrium model is used to quantify pH-dependent reactions. This approach is used here to develop OTEQ, a solute transport model that couples **O**ne-dimensional **T**ransport with **EQ**uilibrium chemistry.

1.2   Applicability

OTEQ is generally applicable to solutes which undergo reactions that are sufficiently fast relative to hydrologic processes (the *"Local Equilibrium Assumption"*; Di Toro, 1976; Rubin, 1983). Although the definition of "sufficiently fast" is highly solute and application dependent, many reactions involving inorganic solutes quickly reach a state of chemical equilibrium. Given a state of chemical equilibrium, inorganic solutes may be modeled using OTEQ's equilibrium approach. This equilibrium approach is facilitated through the use of an existing database that describes chemical equilibria for a wide range of inorganic solutes. In addition, solute reactions not included in the existing database may be added by defining the appropriate mass-action equations and the associated equilibrium constants. As such, OTEQ provides a general framework for the modeling of solutes under the assumption of chemical equilibrium. Despite this generality, most OTEQ applications to date have focused on the transport of metals in streams and small rivers. The remainder of this document is therefore focused on metal transport. Potential model users should note, however, that additional applications are possible.

1.3   Related Reading

Many of the algorithms used within OTEQ are based on the OTIS solute transport model (**O**ne-dimensional **T**ransport with **I**nflow and **S**torage). The reader is therefore encouraged to review the OTIS documentation (Runkel, 1998) in addition to this report. Copies of the OTIS documentation are available from the author or online at http://co.water.usgs.gov/otis. Successful application of OTEQ also requires considerable knowledge of equilibrium chemistry. Model users should review the mathematical treatment of chemical equilibrium problems presented by Morel and Hering (1993) and the software documentation for the U.S. Environmental Protection Agency's MINTEQ program (Allison and others, 1991).

1.4   Report Organization

The remaining sections of this report are as follows. Section 2 provides a description of the theoretical constructs underlying the reactive transport model. This section includes descriptions of the simulated processes, the governing differential equations, and the numerical methods used within the model. Section 3, a User's Guide, presents the input and output requirements of the Fortran computer program. Model parameters, print options, and simulation control variables are detailed in this section. Section 4 presents several applications of the model and includes example input and output files. The final section, a Software Guide (Section 5), describes how to obtain the model, installation procedures, and several programming features.

1.5   Acknowledgments

The groundwork for OTEQ was completed in the late 1980s and early 1990s by Ken Bencala, Briant Kimball, and Diane McKnight, all of the U.S. Geological Survey (USGS). These researchers conducted a series of field-scale experiments to study metal fate and transport. Analysis of these experiments led to the development of what is now known as OTEQ. Bob Broshears (USGS) conducted many of the early OTEQ applications and was instrumental in guiding the initial development effort. The author thanks Ming-Kuo Lee (Auburn University) and David Nimick (USGS) for providing helpful review comments on this manuscript. Assistance with software development and distribution was provided by Zac Vohs (USGS). Support for this work was provided by the USGS Toxic Substances Hydrology Program.

2   Theory

This section describes the theoretical constructs underlying the reactive transport model. Section 2.1 begins with a brief overview of how physical transport and chemical equilibrium are coupled. Section 2.2 provides a derivation of the governing differential equations and a description of the general solution algorithm. The detailed algorithms used to describe the physical and chemical processes are presented in Section 2.3. Section 2.4 concludes the theoretical presentation with additional information on numerical methods, the conceptual stream system, and the treatment of boundary conditions.

2.1   Overview

The reactive transport model is formed by coupling the OTIS solute transport model (Runkel, 1998; Runkel, 2000) with a chemical equilibrium submodel. The resultant model considers a variety of physical and chemical processes including advection, dispersion, transient storage, the transport and deposition of waterborne solid phases, acid-base reactions, complexation, precipitation/dissolution, and sorption. Consideration of these processes provides a general modeling framework for the simulation of solute fate and transport.

Solute Transport Model. The OTIS solute transport model is based on a one-dimensional advection-dispersion equation with additional terms to account for lateral inflow and transient storage (Bencala and Walters, 1983). Transient storage has been noted in many streams, where solutes are temporarily detained in eddies and stagnant zones of water that are stationary relative to the faster moving water near the center of the channel. In addition, portions of the flow enter the hyporheic zone (porous areas within the streambed), where solutes are also detained. Lateral inflow represents additional water entering the main channel as surface inflow, overland flow, interflow, and ground-water discharge. Conservation of mass results in a set of partial differential equations describing the physical transport of multiple solutes.

Equilibrium Submodel. The chemical equilibrium submodel is based on MINTEQ (Allison and others, 1991), an extension of the MINEQL model developed by Westall and others (1976). Given analytical concentrations of the chemical components, MINTEQ computes the distribution of chemical species that exist within a batch reactor at equilibrium. These equilibrium computations include the precipitation and dissolution of solid phases as well as sorption processes. The mass-balance and mass-action equations describing equilibria form a set of nonlinear algebraic equations.

The conceptual and mathematical framework underlying MINTEQ and related models is well documented (Westall and others, 1976; Morel and Hering, 1993; Allison and others, 1991). As a result, only the details essential to the development of OTEQ are given here. Chemical "components" are defined as the fundamental building blocks from which all chemical "species" are derived. Chemical reactions involve two or more components that combine to form a chemical species. In general, components are selected such that (1) the components combine linearly to form every possible species, and (2) no component may be formed as a combination of other components (Westall and others, 1976).[1] A species is simply a chemical entity that is formed by combining chemical components. The chemical equilibrium problem entails solving for the unknown species concentrations at equilibrium. This is accomplished by developing mass-action equations to describe the species-producing reactions and mass-balance equations for the chemical components.

Coupling Transport and Equilibrium Chemistry. Coupling transport with chemical equilibrium results in a simultaneous set of algebraic and partial differential equations. The sequential iteration approach (Yeh and Tripathi, 1989) solves the coupled set of equations by dividing each time step into a "reaction" step and a "transport" step. During the reaction step, the equilibrium submodel is executed for each segment in the stream network. Each segment represents a batch reactor wherein chemical equilibrium is assumed. The equilibrium submodel thus determines the solute mass in dissolved, precipitated, and sorbed forms. Based on this information, a transport step is taken in which the solute transport model physically transports the mobile phases of each solute. Because the transport and reaction steps neglect the coupling of the transport and chemistry, the procedure iterates until a specified level of convergence is achieved.

---

[1]One exception to this general rule is the case of multiple oxidations states. For example, Fe(II) can be formed by combining Fe(III) and an electron — Fe(II), Fe(III), and e⁻ are all components.

2.2   Conceptual Model, Governing Equations, and the Sequential Iteration Method

2.2.1   Model Assumptions

The governing equations and solution algorithms used within the reactive transport model are based on the following assumptions:

- **Chemical Equilibria.** Complexation, precipitation/dissolution, and sorption reactions are in a state of local equilibrium. Under this *"Local Equilibrium Assumption,"* chemical reactions are considered sufficiently fast relative to hydrologic processes (Di Toro, 1976; Rubin, 1983). This assumption allows for the use of the equilibrium submodel described above. One exception to the equilibrium approach is the kinetic limitation placed on sorption/desorption reactions involving the streambed (Section 2.3.3).

- **One-Dimensional Transport.** Solute mass is uniformly distributed over the stream's cross-sectional area such that one-dimensional transport is applicable. Given this assumption, equations are developed for a one-dimensional system that consists of a series of stream segments (control volumes). The physical processes affecting solute mass in each stream segment include advection, dispersion, lateral inflow, transient storage, and settling. All dissolved, precipitated, and sorbed species resident in the water column travel at the same advective velocity.

- **Transient Storage.** The physical process of transient storage is in accordance with the OTIS solute transport model (Runkel, 1998): advection and dispersion are not included in the storage zone, where downstream transport is considered negligible; the exchange of solute mass between the main channel and the storage zone is modeled as a first-order mass transfer process (Bencala and Walters, 1983). Chemical processes in the storage zone are described in Section 2.3.5.

- **Physical Parameters.** All model parameters describing physical processes may be spatially variable. Model parameters describing advection and lateral inflow may be temporally variable; these parameters include the volumetric flow rate, main channel cross-sectional area, lateral inflow rate, and the solute concentration associated with lateral inflow. All other model parameters are temporally constant.

- **Chemical Parameters.** All model parameters associated with the equilibrium submodel are spatially and temporally constant.

- **Mobile and Immobile Phases.** Solute mass for each chemical component is distributed among five distinct phases. The first three phases represent dissolved, precipitated, and sorbed mass that is present in the water column. These three phases are mobile, in that they are subject to transport. The final two phases represent precipitated and sorbed mass that resides on immobile substrate (the streambed or stationary debris) in the stream channel; these phases constitute a thin, immobile layer of solute mass that interacts with the overlying water column.

- **Precipitation.** Dissolved mass in the water column may form precipitates if the solution becomes oversaturated with respect to the defined solid phases. Any precipitated mass initially resides in the water column and is subject to transport, until it settles to the streambed or redissolution occurs. Precipitation occurs in the water column exclusively, and precipitation directly to the immobile bed is excluded. Precipitated mass may accumulate on the bed, however, as transported precipitates are subject to the force of gravity and settle at a rate defined by a settling velocity. The settling rate of a particle is unaffected by other solids in solution (Section 2.3.2).

- **Dissolution.** When the aqueous solution is undersaturated, dissolution occurs preferentially from the water column. All of the precipitate in the water column is allowed to dissolve before precipitate on the bed is considered for dissolution. This assumption is based on the intimate contact between precipitates in the water column and the flowing waters (Section 2.3.2).

- **Sorption.** Dissolved species may sorb to solid phases in the water column or to sorption sites on the streambed. Conversely, sorbed species may desorb from sites in the water column or on the streambed. Additional assumptions relative to sorption are presented in Section 2.3.3.

- **Oxidation/Reduction.** Solute mass may be transferred from one chemical component to another as a result of oxidation/reduction reactions (Section 2.3.4).

2.2.2   Derivation of Governing Equations

With these assumptions, the fundamental equations governing reactive solute transport are derived. Governing equations are formulated in terms of the chemical components defined within the equilibrium submodel.[2] The total component concentration, T, is the sum of the dissolved (C), mobile precipitate (P_w), immobile precipitate (P_b), mobile sorbed (S_w), and immobile sorbed (S_b) phases:

$$T = C + P_w + P_b + S_w + S_b \tag{1}$$

where each phase consists of one or more chemical species. For example, the total concentration in the dissolved phase is given by:

$$C = c + \sum_{i=1}^{M} a_i x_i \tag{2}$$

where

c concentration of the uncomplexed component species (Fe^{3+}, for example) [moles per liter];

x_i concentration of the *ith* complexed species [moles per liter];

a_i stoichiometric coefficient of the component in the *ith* complexed species;

M number of complexed species;

and species concentrations (c and x_i) are provided by equilibrium computations. Similar relationships for the total precipitated ($P = P_w + P_b$) and total sorbed concentrations ($S = S_w + S_b$) are given by Yeh and Tripathi (1989).

A summary of the processes considered for each phase is presented in figure 1, where the system is represented as two compartments. The water column compartment contains the three mobile phases, C, P_w, and S_w. Immobile substrate (the streambed or debris) constitutes the second compartment, containing the two immobile phases, P_b and S_b. Mass transfer between phases is quantified using source/sink terms (f_b, f_w, g_b, g_w; see arrows, fig. 1). The three mobile phases are subject to physical transport, as represented by the transport operator, $L(\ )$. The dissolved phase, C, takes part in precipitation/dissolution and sorption/desorption reactions that occur within the water column (interactions with P_w and S_w; f_w and g_w arrows, fig. 1). The dissolved phase is also affected by dissolution of precipitate from the immobile substrate and by sorption/desorption from immobile sorbents (interactions with P_b and S_b; f_b and g_b arrows, fig. 1). Finally, C may increase or decrease due to external sources and sinks, as denoted by s_{ext} (gas exchange between the atmosphere and the water column, for example). The precipitated and sorbed phases in the water column settle in accordance with the settling velocity, v_1 [LT^{-1}].

Conceptual Surface-Water System

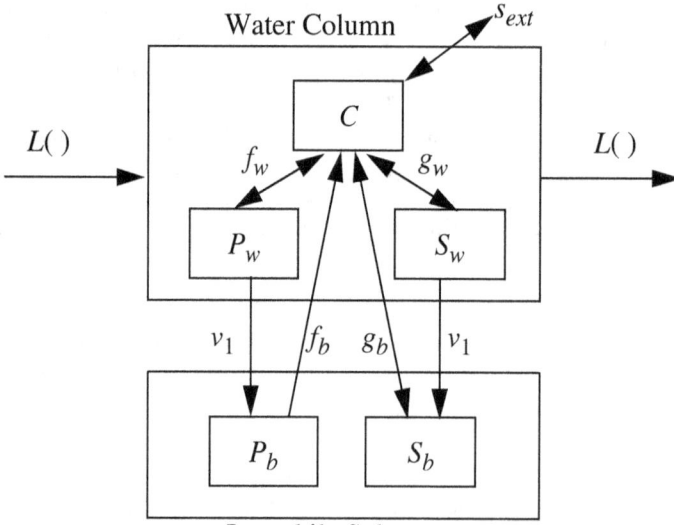

Figure 1.   Conceptual surface-water system used to develop the governing differential equations. The total component concentration consists of dissolved (C), mobile precipitate (P_w), immobile precipitate (P_b), mobile sorbed (S_w), and immobile sorbed (S_b) phases. The dissolved and mobile phases are subject to transport, as denoted by $L(\ )$. Mass transfer between phases is quantified using source/sink terms ($f_b, f_w, g_b, g_w, s_{ext}$) and a settling velocity (v_1) as described in the text.

---

[2]A full listing of chemical components is provided in the MINTEQ documentation (Allison and others, 1991). Examples of chemical components include anions (chloride, sulfate, fluoride), cations (aluminum, ferrous iron, ferric iron), computed quantities (total excess hydrogen), and sorptive surfaces.

A general mass-balance equation for each component is developed by considering the mass associated with each of the five phases within a stream segment (control volume). An equation describing conservation of mass for each component is then developed by summing the equations for the individual phases. In the derivations that follow, the compartments depicted in figure 1 are not treated as separate control volumes, but rather as a single control volume for which a macroscopic mass balance applies (Bird and others, 1960). Note that this approach differs from the approach used in contemporary sediment-water models for toxic substances. These models are often developed for rivers and lakes in which significant volumes of sediment interact with the water column. In this instance, two or more control volumes are used to represent the sediments and the water column. For our purposes, we are concerned with streams where only a thin, immobile layer of precipitated and sorbed mass interacts with the overlying water column. As such, treatment of the system as a single control volume is an appropriate approach.

Mass-balance equations for the five phases are developed below. To simplify the presentation, the precipitate phase for each component consists of a single species. The dissolved and sorbed phases, meanwhile, are not limited by this assumption and may be composed of multiple species. The problem of multiple precipitate species for a single component is revisited in Section 2.3.2. Mass balances for the five phases are given by:

Dissolved Phase

$$\frac{\partial C}{\partial t} = L(C) + f_w + f_b + g_w + g_b + s_{ext} \tag{3}$$

Mobile Precipitate

$$\frac{\partial P_w}{\partial t} = L(P_w) - f_w - \frac{v_1}{d_1}P_w \tag{4}$$

Mobile Sorbate

$$\frac{\partial S_w}{\partial t} = L(S_w) - g_w - \frac{v_1}{d_1}S_w \tag{5}$$

Immobile Precipitate

$$\frac{dP_b}{dt} = \frac{v_1}{d_1}P_w - f_b \tag{6}$$

Immobile Sorbate

$$\frac{dS_b}{dt} = \frac{v_1}{d_1}S_w - g_b \tag{7}$$

where[3]

L transport operator;

f_w source/sink term for precipitation/dissolution from the water column [moles per liter T^{-1}];

f_b source/sink term for dissolution from the immobile substrate [moles per liter T^{-1}];

g_w source/sink term for sorption/desorption from the water column [moles per liter T^{-1}];

g_b source/sink term for sorption/desorption from the immobile substrate [moles per liter T^{-1}];

s_{ext} source/sink term representing external gains and losses [moles per liter T^{-1}];

v_1 main channel settling velocity [LT^{-1}];

d_1 effective settling depth [L] (see Section 2.3.6); and

t time [T].

Given these mass-balance equations, several comments are in order. First, the source/sink terms (f_w, f_b, g_w, g_b, s_{ext}) are implicit functions that are dependent on the solution of the nonlinear algebraic equations describing chemical equilibria (source/sink terms are not explicitly provided by the equilibrium submodel; algorithms to develop these terms are provided in Section 2.3). Second, the external source/sink term (s_{ext}) represents mass that is added to (or lost from) the system due to the presence of a source (or sink) that is external to the system; unlike the other source/sink terms, s_{ext} does not represent mass transfer between the five phases. For example, the equilibrium submodel may be used to describe an aqueous system that is in equilibrium with atmospheric CO_2. This use of the equilibrium submodel results in a gain (transfer from the atmosphere to the dissolved phase) or loss (degassing) of mass due to an external source/sink. Another example is the specification of an infinite solid (Allison and others, 1991). Additional details on s_{ext} are given in Section 2.2.3 and by Runkel (1993). Finally, the transport operator is defined in terms of the transient storage model (Bencala and Walters, 1983; Runkel, 1998):

---

[3]The fundamental units of Length [L] and Time [T] are used throughout this section. Specific units are introduced in Section 3.

$$L(\hat{C}) \;=\; -\frac{Q}{A}\frac{\partial \hat{C}}{\partial x} + \frac{1}{A}\frac{\partial}{\partial x}\!\left(AD\frac{\partial \hat{C}}{\partial x}\right) + \frac{q_{LIN}}{A}(\hat{C}_L - \hat{C}) + \alpha(\hat{C}_S - \hat{C}) \tag{8}$$

where

A	main channel cross-sectional area [L^2];
$\hat{C}$	main channel concentration of an arbitrary phase [moles per liter];
$\hat{C}_L$	lateral inflow concentration of the arbitrary phase [moles per liter];
$\hat{C}_S$	storage zone concentration of an arbitrary phase [moles per liter];
D	dispersion coefficient [L^2T^{-1}];
Q	volumetric flow rate [L^3T^{-1}];
q_{LIN}	lateral inflow rate [$L^3T^{-1}L^{-1}$];
x	distance [L]; and
α	storage zone exchange coefficient [T^{-1}].

Use of the transient storage approach introduces an additional set of mass-balance equations for the storage zone concentrations, $\hat{C}_S$. The storage zone equations are discussed in Section 2.3.5. Nomenclature is introduced here to distinguish between parameters that apply to the main channel and those that apply to the storage zone. Parameters v_1 and d_1 contain the subscript "1" to denote the main channel; the subscript "2" denotes the corresponding parameters in the storage zone.

The mass-balance equation for the total component concentration, T, is obtained by summing the mass-balance equations for the five individual phases. This yields:

$$\frac{\partial T}{\partial t} \;=\; L(C + P_w + S_w) + s_{ext}. \tag{9}$$

2.2.3   A General Solution Scheme based on Sequential Iteration

The basic problem is as follows. Given the mass-balance equations developed in Section 2.2.2, how can the equilibrium submodel be used to determine the component concentrations in the various phases? For the ground-water systems described by Yeh and Tripathi (1989), the total component concentration consists of three distinct phases (dissolved, precipitated, and sorbed); total concentrations in each of these three phases are readily available as output from the equilibrium submodel. For the present case, use of the equilibrium submodel is confounded by the presence of five phases. The additional phases are due to division of precipitated and sorbed mass into mobile and immobile fractions. Fortunately, the equilibrium submodel allows for the definition of multiple sorptive surfaces, so that separate surfaces may be defined for the two sorbed fractions. This feature allows one to differentiate between the mobile and immobile sorbate concentrations. For the case of precipitation/dissolution, such a feature is unavailable, and only a single value reflecting the total amount of precipitate is provided. An algorithm to determine the amount of the mobile and immobile precipitate is therefore required.

The solution technique presented here uses total component concentration, T, as the primary variable. A differential equation for T is presented as equation 9. This equation is analogous to the explicit form of the ground-water equation (Yeh and Tripathi, 1989). Here an implicit form is developed by combining equations 1 and 9:

$$\frac{\partial T}{\partial t} \;=\; L(T) - L(S_b + P_b) + s_{ext} \tag{10}$$

where equation 10 is formulated such that the waterborne phases are eliminated. Inspection of equation 10 reveals that T is a function of P_b and S_b. Equations for these phases are given by:

$$\frac{dP_b}{dt} \;=\; \frac{v_1}{d_1}(P - P_b) \;-\; f_b \tag{11}$$

$$\frac{dS_b}{dt} \;=\; \frac{v_1}{d_1}(S - S_b) \;-\; g_b \tag{12}$$

where P and S are the total precipitated ($=P_w + P_b$) and total sorbed ($=S_w + S_b$) concentrations (P_w and S_w have been eliminated from eqs. 6 and 7).

The equation set governing the problem consists of three partial differential equations (for T, P_b, and S_b) for each component and the set of algebraic equations representing chemical equilibria. This equation set is solved using a Crank-Nicolson approximation of the governing differential equations and the sequential iteration approach. Presentation of the solution technique requires additional nomenclature. Let n denote an initial time and $n+1$ denote an advanced time; time n is the previous time at which the state of the system is known, and time $n+1$ is the current time for which a solution is desired. In addition, k is a counter used to denote the iteration number. Finally, a caret is used to indicate that a given quantity is an estimate. For example, $\hat{C}^{n+1,k}$ is an estimate of the dissolved concentration for the current iteration at the advanced time level. Additional details on the numerical solution scheme and the Crank-Nicolson method are provided in Section 2.4 and by Runkel (1998).

The goal of sequential iteration is to solve the set of partial differential equations describing transport. In general, there is one equation in the form of equation 10 for each chemical component. Values of the state variables at the initial and advanced time levels are needed to solve for the total component concentrations at the advanced time level (T^{n+1}) using Crank-Nicolson. The state variables at time level n are available from the previous time step, while estimates of the state variables must be made for time level $n+1$. Specifically, estimates of P, P_b, S, and S_b are needed, as well as the source/sink terms f_b, g_b, and s_{ext}. As shown below, P and S are provided directly by the equilibrium submodel, P_b and S_b are provided via equations 11 and 12, and the source/sink terms are developed algorithmically. Iteration is required because the values based on the equilibrium calculations are only estimates of the variables at the advanced time level. Solution of the reactive transport problem consists of four steps: initialization, equilibrium calculations, transport calculations, and convergence testing (fig. 2).

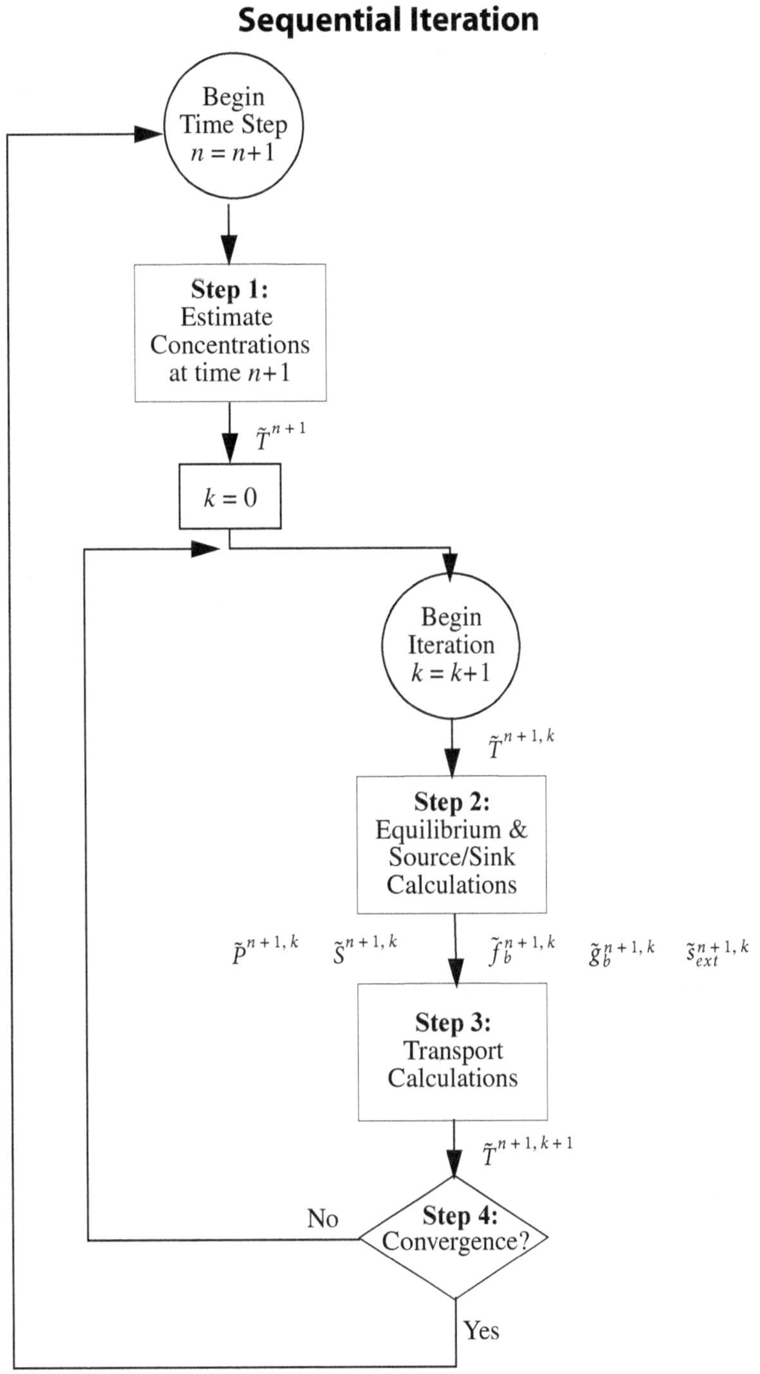

Sequential Iteration

Figure 2.   The sequential iteration approach for the reactive surface-water model.

Step 1: Initialization

As each time step begins, the total component concentration at the advanced time level is estimated for each component ($\tilde{T}^{n+1}$). These concentrations are used as input to Step 2. Values for $\tilde{T}^{n+1}$ are obtained using values from the previous time step, by simply setting $\tilde{T}^{n+1}$ equal to T^n. This estimation procedure is only completed at the beginning of each time step, prior to completing the equilibrium and transport steps within the first iteration. Refined estimates are obtained within the iterative loop, as described below.

Step 2: Equilibrium Calculations

Step 2 begins the iterative loop (fig. 2). For each chemical component, the estimate of the total component concentration ($\tilde{T}^{n+1}$) is checked to ensure that it is a valid concentration. If $\tilde{T}^{n+1}$ is less than a prescribed minimum value (1×10^{-20} moles per liter), $\tilde{T}^{n+1}$ is reset to the prescribed minimum. This procedure ensures that zero or negative component concentrations are not passed to the equilibrium submodel. Final estimates are then input to the equilibrium submodel where the concentrations of the chemical species are computed. The concentrations of the individual species are summed to yield the total component mass in the dissolved, precipitated, and sorbed phases (C, P, and S). These quantities are used to compute the dissolution source/sink (f_b) and the sorption/desorption source/sink (g_b) as described in Sections 2.3.2 and 2.3.3.

Gain or loss via external sources and sinks is quantified by comparing the total component concentrations before and after the equilibrium calculations. As before, the total component concentrations used as input to the equilibrium submodel are denoted as $\tilde{T}^{n+1,k}$. The total component concentrations after equilibration are given by the sum of the dissolved, precipitated, and sorbed phases. The external source/sink term for each component is therefore computed by:

$$\tilde{s}_{ext}^{n+1,k} = \frac{\tilde{C}^{n+1,k} + \tilde{P}^{n+1,k} + \tilde{S}^{n+1,k} - \tilde{T}^{n+1,k}}{\Delta t} \tag{13}$$

where Δt is the integration time step [T].

Step 3: Transport Calculations

Given estimates of f_b, g_b, and s_{ext} at time $n+1$, the equations describing transport and settling are now solved. Equations 11 and 12 are first solved for the immobile precipitate and sorbed phases. The estimates of P_b, S_b, and s_{ext} are then used in conjunction with the variables from time n to solve equation 10 using the Crank-Nicolson method. The value of $\tilde{T}^{n+1,k+1}$ so obtained represents one of two states. If the solution has converged (Step 4), this concentration represents the final solution to the reactive transport problem for the current time step. If convergence is not obtained, $\tilde{T}^{n+1,k+1}$ is a refined estimate of the component concentrations used as input for Step 2 in the next iteration.

Step 4: Convergence Test

In Step 2, phase concentrations at the advanced time level ($n+1$) are determined via chemical equilibrium calculations. These calculations are based on estimates of the total component concentrations at the advance time level ($\tilde{T}^{n+1,k}$). For the first iteration, these estimates are based on the previous value of T, as described under Step 1. For subsequent iterations, the estimates are based on the solution of the transport equations (Step 3) from the previous iteration. As the iterative technique progresses, these estimates should approach T^{n+1}, and the solution converges. The algorithm therefore requires some objective mechanism whereby a test for convergence is performed. This convergence test is given by:

$$\left| \frac{\tilde{T}^{n+1,k+1} - \tilde{T}^{n+1,k}}{\tilde{T}^{n+1,k+1}} \right| < \sigma_e \tag{14}$$

where σ_e is a relative error tolerance. If equation 14 holds for all components in all segments, the solution has converged and a new time step is initiated. If the left-hand side of equation 14 is greater than σ_e for any component in any segment, the solution has not converged and another iteration is required.

2.3   Process Formulation

A general algorithm for the solution of the reactive transport problem is presented in the foregoing section. This section presents a detailed description of the solution techniques used to implement specific processes within the model. These processes include pH, precipitation/dissolution, sorption, oxidation/reduction, transient storage, and the settling of solid phases.

2.3.1   pH

Models of chemical equilibria generally use one of two approaches for the calculation of pH. Under the electroneutrality approach, a charge balance equation is used to determine the aqueous concentration of H^+. This approach is implemented within the PHREEQC equilibrium model (Parkhurst and Appelo, 1999). A second approach, based on the proton condition (Morel and Morgan, 1972), is used within the reactive transport model. Under this approach, a mass-balance equation is written for the "excess" hydrogen ions in solution. The proton condition approach is advantageous in that excess hydrogen may be defined as an aqueous component that is subject to transport (eq. 10); no special treatment of acid-base chemistry is required (Yeh and Tripathi, 1991). Within the equilibrium submodel, the proton condition is given by:

$$T_H = \sum H^+ \ species - \sum OH^- \ species \qquad (15)$$

where T_H is the total component concentration for excess hydrogen. Additional details on the proton condition and the simulation of pH are provided in Sections 4.2–4.5.

2.3.2   Precipitation/Dissolution

Step 3 of the sequential iteration procedure (Section 2.2.3) uses an estimate of the dissolution source/sink term to solve the immobile precipitate equation (eq. 11). As shown here, information obtained from the equilibrium submodel may be used to estimate f_b for each stream segment. To begin, consider the change in total precipitate, P, from one time step to the next. Re-examining the differential equations derived for each phase, the change in P with time is given by the sum of equations 4 and 6:

$$\frac{\partial P}{\partial t} = L(P_w) - f_w - f_b. \qquad (16)$$

Estimation of f_b from equation 16 is based on several simplifying assumptions. Two cases are of interest. First, the total amount of precipitate may increase ($\partial P/\partial t > 0$). This increase may be due to precipitation ($f_w < 0$) and(or) transport of mobile precipitate [$L(P_w) > 0$; eq. 8]. If precipitation is occurring, dissolution is not possible and f_b is zero. The total amount of precipitate may also increase if the gain due to transport is greater than the loss due to dissolution [$L(P_w) > 0$ and $L(P_w) > f_w + f_b$]. For this latter situation, f_b is also zero, as the gain due to transport indicates the presence of P_w. (Recall that dissolution is assumed to occur preferentially from the water column, such that a nonzero P_w implies an f_b of zero). The second case is when the total amount of precipitate decreases ($\partial P/\partial t < 0$). This decrease may be due to dissolution ($f_w > 0$, $f_b \geq 0$) and(or) transport of mobile precipitate [$L(P_w) < 0$]. At a given location, dissolution from the bed occurs ($f_b > 0$) only after the supply of mobile precipitate (P_w) has been exhausted. This observation is used to eliminate $L(P_w)$ from equation 16. Heuristics may then be used to differentiate between f_w and f_b. This final assumption is not entirely valid, as situations may arise in which $L(P_w)$ is significant. These situations do not present a problem, however, as the presence of precipitate in the water column results in an f_b of zero.

An algorithm for determining f_b is presented in figure 3. Using estimates of the total component concentrations, the equilibrium submodel is called to compute the total amount of precipitate present ($\tilde{P}^{n+1,k}$). If the total amount of precipitate at time $n+1$ exceeds that for time n ($\tilde{P}^{n+1,k} > P^n$), the net amount of precipitate has increased, indicating that precipitation has occurred. For the case of precipitation, f_b equals zero. If, however, the net amount of precipitate had decreased ($\tilde{P}^{n+1,k} < P^n$), dissolution has occurred and two situations are possible. First, if net dissolution (defined as $P^n - \tilde{P}^{n+1,k}$) is less than the amount of precipitate present in the water column ($P_w{}^n$), all of the dissolved mass is taken from the mobile phase and no dissolution occurs from the bed. Here again f_b equals zero. Second, if net dissolution is not accounted for by mass residing in the mobile phase, mass has dissolved from the immobile phase. In this case, f_b is given by:

$$\tilde{f}_b^{n+1,k} = \frac{Net \ Dissolution \ - \ P_w^n}{\Delta t}. \qquad (17)$$

Dissolution Source/Sink

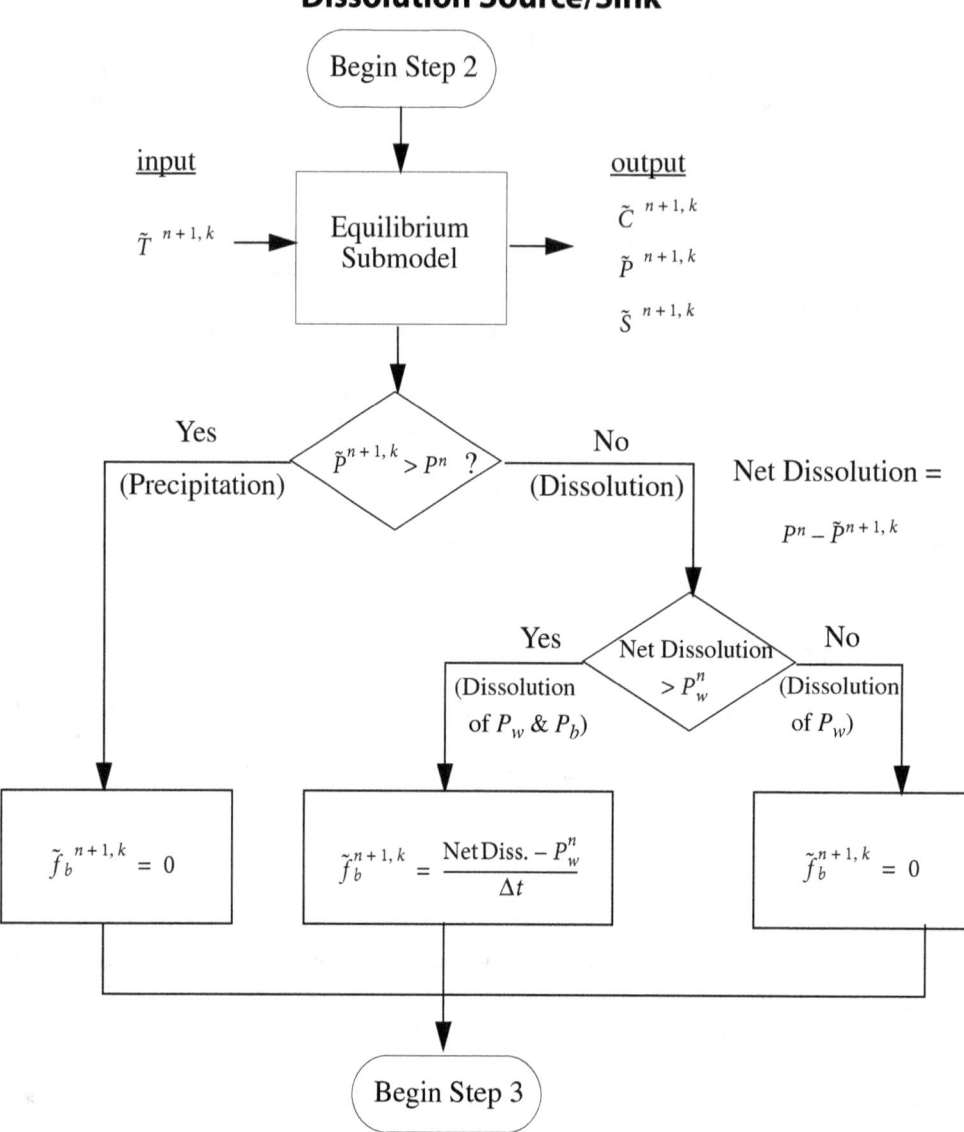

Figure 3.   Computation of the dissolution source/sink term (f_b).

The Problem of Multiple Precipitates. The equations developed above are based on the assumption that the precipitate phase consists of a single species. In this section the introduction of multiple precipitate species for a single component is examined. Note that equation 6 is developed by considering the change in component mass due to settling and dissolution from the bed. When more than one precipitate is present for a given component, the settling and dissolution terms in equation 6 are incorrect, as precipitate species may settle at different velocities (hence v_1 cannot be specified on a component basis), and dissolution from the immobile phase is species-specific. To consider multiple precipitates correctly, mass-balance equations are developed for each precipitated species:

$$\frac{dp_{bm}}{dt} = \frac{v_{1m}}{d_1}(p_m - p_{bm}) - f_{bm} \tag{18}$$

where

- p_{bm} immobile precipitate concentration for precipitated species m [moles per liter];
- p_m total precipitate concentration for precipitated species m [moles per liter];
- v_{1m} settling velocity for precipitated species m [LT^{-1}]; and
- f_{bm} source/sink term for dissolution of immobile precipitated species m [moles per liter T^{-1}].

Solution of the reactive solute transport problem now requires a modified approach. Step 3 of the sequential iteration procedure (Section 2.2.3) is modified as follows. First, rather than solving equation 11, equation 18 is solved for each precipitated species associated with a given component. To solve equation 18, the amount of precipitate for species m ($\tilde{p}_m^{n+1,k}$) is obtained from the equilibrium submodel. The source/sink term for species m ($\tilde{f}_{bm}^{n+1,k}$) is also required and is developed using a procedure analogous to that for $\tilde{f}_b^{n+1,k}$. After solving the m equations to obtain the immobile precipitate concentrations, the total component concentration of immobile precipitate is given by:

$$\tilde{P}_b^{n+1,k} = \sum_{m=1}^{np} a_m \tilde{p}_{bm}^{n+1,k} \qquad (19)$$

where a_m is the stoichiometric coefficient of the component in the *mth* precipitated species, and np is the number of solid precipitate species for the current component. The solution of the governing equation (eq. 10) then proceeds as before.

2.3.3   Sorption

Step 3 of the sequential iteration procedure (Section 2.2.3) uses an estimate of the sorption/desorption source/sink term to solve the immobile sorbate equation (eq. 12). As shown here, information obtained from the equilibrium submodel may be used to estimate g_b. Mathematical descriptions of sorption range from simple distribution-coefficient approaches to more complex representations based on electro-chemical theory. Because distribution-coefficient approaches neglect electrostatic effects, their use is limited in metal-contaminated waters where the primary sorbents are hydrous metal oxides that have charged surfaces. The reactive transport model therefore estimates g_b using the generalized two-layer model (Dzombak and Morel, 1990), a surface complexation model that explicitly considers the effects of pH and ionic strength on surface charge.

Generalized Two-Layer Model (GTLM). This section describes the generalized two-layer model (Dzombak and Morel, 1990) as implemented within the equilibrium submodel. As with other surface complexation models, the generalized two-layer model defines sorption reactions in terms of mass law equations that govern the concentrations of sorbate, sorbent, and surface sites at equilibrium. The equilibrium constant associated with a given mass-action equation is the product of an intrinsic term representing the chemical free energy of site binding and a second term representing the coulombic free energy of binding due to the electrostatically charged surface. The coulombic term acts as a surface activity coefficient that accounts for the work required to move ions from the surface layer to the bulk solution.

Because the coulombic term varies as a function of surface charge and potential, sorption mass law equations must be rearranged and expressed in terms of intrinsic surface complexation constants. For example, consider sorption of a divalent cation:

$$SOH + M^{2+} \leftrightarrow SOM^+ + H^+ \qquad (20)$$

where M^{2+} is a divalent cation, H^+ is a hydrogen ion, SOH is an uncharged surface hydroxyl group, and SOM^+ is a positively charged surface species. The corresponding mass-action equation is:

$$K = \frac{\{SOM^+\}\{H^+\}}{\{SOH\}\{M^{2+}\}} \qquad (21)$$

where K is the equilibrium constant and { } denotes chemical activity. Expressing K as the product of the intrinsic and coulombic terms and rearranging yields:

$$K^{int} = \frac{\{SOM^+\}\{H^+\}}{\{SOH\}\{M^{2+}\}\exp\left(\dfrac{-\Psi F}{RT_a}\right)} \qquad (22)$$

where K^{int} is the intrinsic surface complexation constant, $\exp(-\Psi F/RT_a)$ is the coulombic correction factor, Ψ is surface potential [volts], F is the Faraday constant [96,485 coulomb mole^{-1}], R is the molar gas constant [8.314 joules mole^{-1} K^{-1}], and T_a is absolute temperature [K].

Solution of a chemical equilibrium problem that includes equations such as 22 requires introduction of a dummy chemical component to account for the coulombic correction factor and a definition of surface potential. Under electrical double layer theory, surface charge is balanced by a diffuse layer of counter charges in solution; the relationship between surface charge and surface potential is defined by Guoy-Chapman theory (Dzombak and Morel, 1990):

$$\sigma = \sqrt{8RT_a \varepsilon \varepsilon_0 c_e 10^3}\,\sinh\left(\frac{Z\Psi F}{2RT_a}\right) \qquad (23)$$

where σ is net surface charge density [coulomb meter^{-2}], ε is the dielectric constant of water, ε_o is the permittivity of free space [$8.876{\times}10^{-12}$ coulomb volt^{-1} meter^{-1}], c_e is the molar electrolyte concentration, and Z is the valence of a symmetrical electrolyte. Given equation 23, the total component concentration for the coulombic correction factor [moles per liter] is given by:

$$T_{CC} = \sigma\frac{S_A S_C}{F} \qquad (24)$$

where $CC = \exp(-\Psi F/RT_a)$, S_A is specific surface area [meter^2 per gram sorbent], and S_C is solid concentration [gram sorbent per liter].

Solution of the chemical equilibrium problem also requires specification of components that represent the sorptive surface. A central part of the generalized two-layer model is the postulation that each sorptive surface has two types of sites for cation binding. The first type, the high-affinity site, is generally less prevalent than the second site type but has a stronger binding potential. A second low-affinity site is in greater abundance but has weaker binding potential. Due to presence of two site types, two chemical components are introduced for each sorptive surface. Total component concentration [moles of sites per liter] for each site type is given by:

$$T_{SOH} = \frac{N_s S_C}{M} \qquad (25)$$

where N_S is the site density [moles of sites per mole sorbent] and M is the molecular weight of the sorbent [gram sorbent per mole sorbent].

GTLM within the Reactive Transport Model. The reactive transport model is formulated such that sorption may occur onto static and dynamic sorptive surfaces. Static sorptive surfaces are those for which the concentration of sorptive solid (S_C) does not change in time. Conversely, dynamic sorptive surfaces are those for which the concentration of the sorptive solid is time variable. An example of a static sorptive surface is a streambed armored with hydrous iron oxides (Broshears and others, 1996). In this case the number of sites available for sorption reactions is relatively constant throughout the time period of interest. Other situations may arise in which the number of sites changes in time, and sorption to a dynamic surface is applicable. Such is the case when hydrous metal oxides form in the water column as a result of precipitation reactions. Within the model, these precipitates are defined as dynamic sorptive surfaces.

To further classify the sorption reactions, it is useful to subdivide the sorptive surfaces into three "pools." Pool 1 consists of the static sorptive surface; Pool 2 consists of the dynamic surfaces present in the water column, that is, associated with the water-borne precipitates; and Pool 3 consists of dynamic surfaces that were initially present in Pool 2 but have settled to the streambed during the course of the simulation. The sorbed concentrations associated with Pools 1, 2, and 3 are denoted by S_1, S_2, and S_3, respectively. Additional assumptions underlying the use of GTLM within the reactive transport model are as follows:

- Sorption reactions adhere to the generalized two-layer model as defined by Dzombak and Morel (1990).
- A static surface and(or) a dynamic surface may be defined. Each surface may have high- and low-affinity sites. A dynamic surface is distributed between Pools 2 and 3 as defined above.
- Specific surface area (S_A) and sorbent molecular weight (M) are specified for each surface. Site density (N_S) is specified for each site type on each sorptive surface. Sorbent properties (S_A, M, N_S) are spatially and temporally constant.

Given these assumptions, three cases are possible: (1) sorption to a static surface, (2) sorption to a dynamic surface, and (3) sorption to static and dynamic surfaces. As shown below, these cases differ with respect to how equation 12 is solved, how the equilibrium submodel is used, and how kinetic limitations are imposed.

Sorption to a Static Surface

A conceptual diagram depicting sorption to a static surface is given as figure 4. When sorption occurs, mass is transferred from the dissolved phase, C, to the sorbed phase associated with Pool 1, S_1. When only a static surface is considered, S_1 is equivalent to the immobile sorbed phase, S_b. During desorption, mass is transferred from S_1 to C. The rate of sorption/desorption is governed by the kinetic parameter, Γ. Sorption to a static surface requires two additional assumptions:

- The solid concentration (S_C) does not change in time. The sorptive solid is attached to immobile substrate (the streambed or debris) and is therefore not subject to downstream transport.
- The solid concentration is allowed to vary spatially on a reach-specific basis.

Sorption to a Static Surface

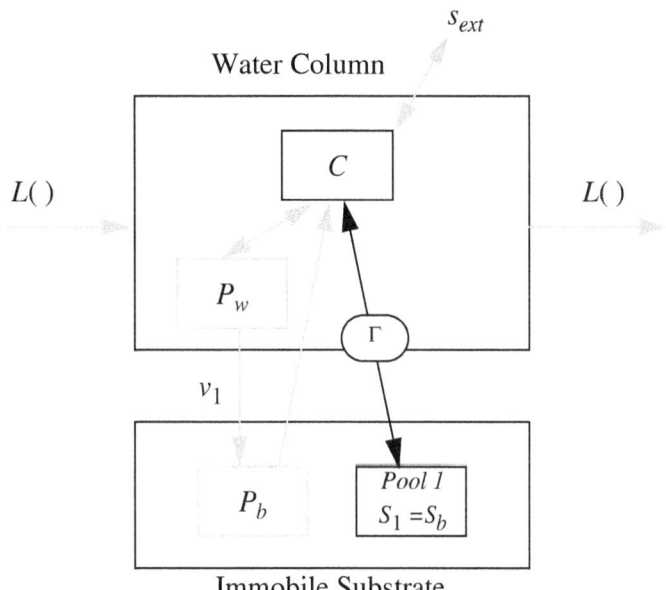

Figure 4. Conceptual surface-water system for sorption to a static surface.

The primary task is to solve equation 12 for the concentration of the component sorbed to the static surface, S_b. For static surfaces, the settling term in equation 12 drops out, yielding:

$$\frac{dS_b}{dt} = -g_b \qquad (26)$$

where g_b is estimated using output from the equilibrium submodel. Two cases of sorption to static surfaces are now considered: equilibrium sorption and kinetically limited sorption.

Equilibrium Sorption. The first step in modeling equilibrium sorption is to determine the total component concentrations (T^{n+1}) used as input to the chemical equilibrium submodel. For the chemical components, T^{n+1} corresponds to the solution of equation 10 from the previous sequential iteration. Total component concentrations for the coulombic correction factor and the high- and low-affinity sites on the static sorptive surface are given by equations 24 and 25. The sorbent concentration (S_C) used in equation 24 is equal to the temporally constant, spatially variable value assigned at the beginning of the simulation. Specific values of S_C, S_A, N_S, and M are shown in table 26 (Section 3.4.3), Section 4.3, and Section 4.5.

As shown in figure 5a, the equilibrium submodel determines the total component mass in the dissolved, precipitated, and sorbed phases $(C^{n+1}, P^{n+1}, S^{n+1})$. The sorption source/sink term is then calculated based on the change in sorbed concentration during the current time step:

$$g_b = \frac{S^n - S^{n+1}}{\Delta t} \qquad (27)$$

Given this definition of g_b, equation 26 may now be solved using a forward time difference ($dS_b = S_b^{n+1} - S_b^n$) and the fact that S_w equals zero when only a static surface is considered ($S^n = S_b^n$). This yields:

$$S_b^{n+1} = S^{n+1}. \qquad (28)$$

As shown by equation 28, the submodel provides the exact quantity needed for the solution, S^{n+1}. As such, the model uses the equality given by equation 28, rather than a formal solution of equation 12.

Equilibrium Submodel: Static Surface

a. Equilibrium Sorption

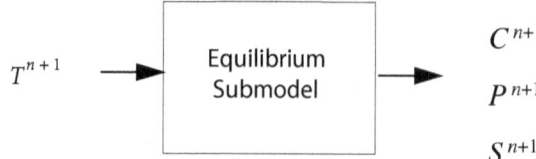

b. Kinetically Limited Sorption

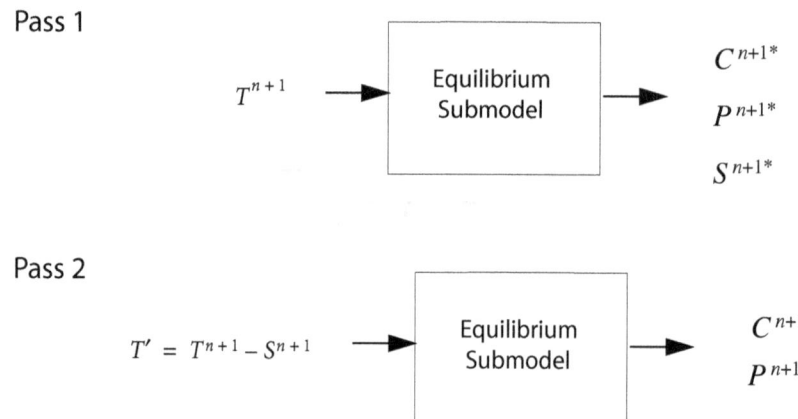

Figure 5. Use of the equilibrium submodel for a static surface: (a) equilibrium and (b) kinetically limited sorption.

Kinetically Limited Sorption. Under the equilibrium approach, the sorbed concentration is taken directly from the equilibrium submodel. Here we employ a pseudo-kinetic approach in which only a fraction of the change in sorbed concentration is considered. This kinetic limitation is designed to model cases where only a portion of the mass in the water column comes in contact with the static surface (the streambed). Figure 5b depicts use of the equilibrium submodel for kinetically limited sorption. Computations during Pass 1 are very similar to the computations described for equilibrium sorption; given T^{n+1}, the submodel determines the total component mass in the dissolved, precipitated, and sorbed phases (C^{n+1*}, P^{n+1*}, and S^{n+1*}, where * denotes the equilibrium concentration in the absence of a kinetic limitation). The sorption source/sink term is now equal to a fraction of the change in sorbed concentration:

$$g_b = \Gamma \left(\frac{S^n - S^{n+1}}{\Delta t} \right)^*$$

(29)

where Γ is the fraction of the equilibrium quantity that is allowed to sorb/desorb during the current time step. Solving equation 26 using a forward time difference yields:

$$S_b^{n+1} = S_b^n + \Gamma(S^{n+1*} - S^n)$$

(30)

During Pass 1, output from the equilibrium submodel reflects solution chemistry under the assumption of chemical equilibrium. The phase concentrations from the submodel (C, S, P) and the solution pH therefore do not include the effects of the kinetic limitation. To incorporate these effects, the total sorbed concentration is set equal to the kinetically limited concentration ($S^{n+1} = S_b^{n+1}$), and Pass 2 is initiated. During Pass 2, sorption reactions are not considered and the total component concentrations are revised to eliminate sorbed mass:

$$T' = T^{n+1} - S^{n+1}.$$

(31)

Given T', the equilibrium submodel provides the corrected values of C, P, and pH.

Sorption to a Dynamic Surface

Reactions between the dissolved phase, C, and the dynamic surface in Pools 2 and 3 are depicted in figure 6. Dissolved ions may sorb to the waterborne precipitates in Pool 2, thereby increasing the sorbed concentration associated with Pool 2, S_2. S_2 is equivalent to the mobile sorbed phase (S_w) and is therefore subject to downstream transport and settling. After the reaction occurs, sorbed mass may settle, increasing S_3. Alternatively, desorption may occur from Pool 2, returning mass to the dissolved phase. Sorption/desorption reactions also transfer mass between the dissolved phase and Pool 3. When only a dynamic surface is considered, S_3 is equivalent to the immobile sorbed phase, S_b. Assumptions unique to the dynamic surface are as follows:

- In Pool 2, the concentration of the sorptive solid (S_C) varies in time and space as a function of the mobile precipitate concentration of a specified hydrous metal oxide (P_w^{Me}). The dynamic surface and the associated sorbate reside in the water column and are subject to transport and settling.

- Pools 2 and 3 are modeled using a single surface defined in the submodel. Sorptive solids in Pools 2 and 3 therefore have identical specific surface areas (S_A), molecular weights (M), and site densities (N_S). Mass is apportioned between S_2 and S_3 based on each pool's contribution to S_C.

- In Pool 3, the concentration of the sorptive solid (S_C) varies in time and space as a function of the immobile precipitate concentration of a specified hydrous metal oxide (P_b^{Me}). The dynamic surface and the associated sorbate concentration is attached to immobile substrate (the streambed or debris) and is therefore not subject to downstream transport.

- Sorption/desorption reactions in Pools 2 and 3 are in local equilibrium.

Sorption to a Dynamic Surface

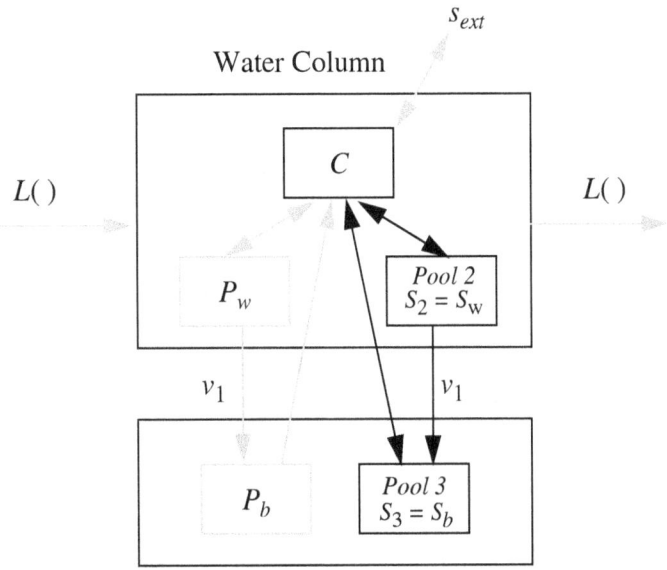

Figure 6. Conceptual surface-water system for sorption to a dynamic surface.

Given these assumptions, the task is to solve equation 12 for the concentration of each component sorbed to the dynamic surface in Pool 3, S_b. As with a static surface, a formal solution to equation 12 is not required; all of the sorbed concentrations may be obtained directly from the equilibrium submodel. The settling term in equation 12 is considered indirectly, in that the settling of sorbed mass is reflected in the settling term for the hydrous metal oxide (sorbed mass settles at the same rate as the precipitate that makes up the dynamic surface, as given by the governing equation for P_b).

Use of the equilibrium submodel for the dynamic surface is nearly identical to that for the static surface under equilibrium conditions (fig. 5a). The only difference is computation of the total component concentrations for the coulombic correction factor (eq. 24) and the surface components (high- and low-affinity sites, eq. 25); the solid concentration (S_C) is now calculated based on the concentration of the precipitate defined as the dynamic surface:

$$S_C = M P^{Me} \tag{32}$$

where P^{Me} is the total precipitate concentration for the specified hydrous metal oxide. As with the static surface under equilibrium conditions, the submodel provides values for C^{n+1}, P^{n+1}, and S^{n+1}. In this case, S^{n+1} is the total concentration sorbed to the dynamic surface. The amount of S^{n+1} on the streambed is determined by considering the fraction of the dynamic surface (precipitate) on the streambed:

$$S_b^{n+1} = \frac{P_b^{Me}}{P^{Me}} S^{n+1} \tag{33}$$

Sorption to Static and Dynamic Surfaces

Concurrent simulation of sorption to static and dynamic surfaces is depicted in figure 7, where interactions between the dissolved phase and the various sorbent pools are shown. As before, sorption/desorption reactions for the static surface may be kinetically limited, whereas sorption/desorption reactions for the dynamic surface are in local equilibrium.

Sorption to Static and Dynamic Surfaces

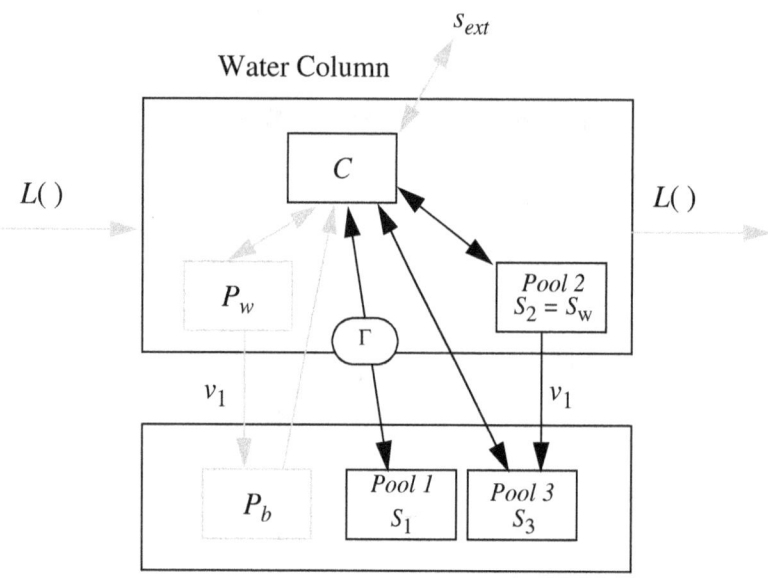

Figure 7.   Conceptual surface-water system for sorption to static and dynamic surfaces.

Equilibrium Sorption. Use of the submodel for equilibrium sorption is shown in figure 8a. Computation of total component concentrations for the coulombic correction factor and the surface components for the static and dynamic surfaces is as described above (eqs. 24 and 25). The submodel provides values of C^{n+1}, P^{n+1}, S_1^{n+1} (static surface, Pool 1), and $S_{2,3}^{n+1}$ (dynamic surface, Pools 2 and 3). The total sorbed concentration is the sum of the concentrations sorbed to the static and dynamic surfaces:

$$S^{n+1} = S_1^{n+1} + S_{2,3}^{n+1} \tag{34}$$

whereas the immobile sorbed concentration is the sum of the concentrations in Pools 1 and 3:

$$S_b^{n+1} = S_1^{n+1} + \frac{P_b^{Me}}{P^{Me}} S_{2,3}^{n+1} . \tag{35}$$

Kinetically Limited Sorption. Use of the submodel for kinetically limited sorption is shown in figure 8b. Pass 1 provides values of C^{n+1*}, P^{n+1*}, S_1^{n+1*}, and $S_{2,3}^{n+1*}$. The kinetically limited concentration sorbed to Pool 1 is given by:

$$S_1^{n+1} = S_1^n + \Gamma(S_1^{n+1} - {}^*S_1^n).$$

(36)

As with a static surface, phase concentrations from Pass 1 and solution pH do not reﬂect the kinetic limitation; corrected values of C, P, $S_{2,3}$, and pH are determined during Pass 2 (ﬁg. 8b), where:

$$T' = T^{n+1} - S_1^{n+1}.$$

(37)

In Pass 2, sorption to the static surface is not considered, as the kinetically limited concentration for Pool 1 has been computed in equation 36. Total sorbed and immobile sorbed concentrations are given by equations 34 and 35.

Equilibrium Submodel: Static and Dynamic Surfaces

a. Equilibrium Sorption

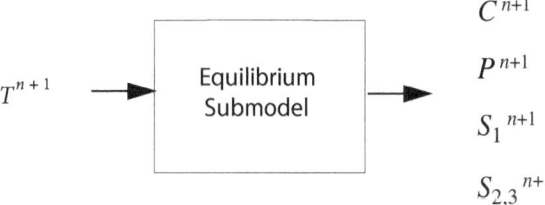

b. Kinetically Limited Sorption

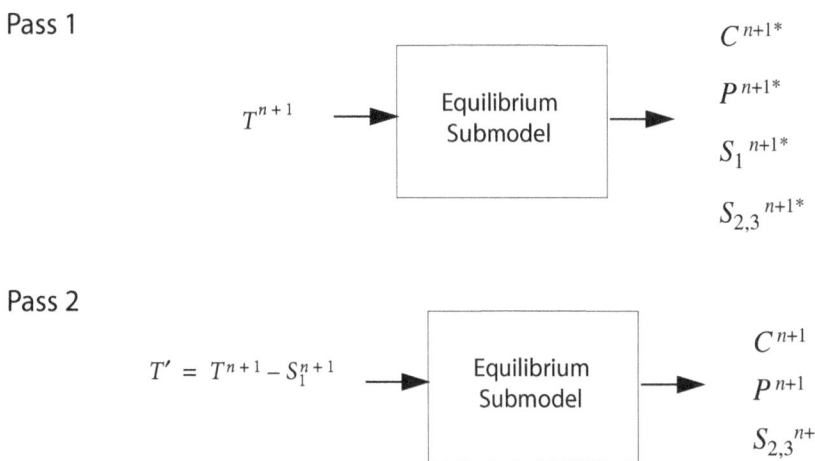

Figure 8.   Use of the equilibrium submodel for static and dynamic surfaces: (a) equilibrium and (b) kinetically limited sorption.

2.3.4   Oxidation/Reduction

Oxidation/reduction reactions play an important role in determining solute concentrations in natural systems. Ferrous and ferric iron concentrations, for example, are often controlled by photoreduction and microbial oxidation (McKnight and others, 1988; Kimball and others, 1992). Previous investigators have used a fixed fraction approach to model the oxidation/reduction of dissolved iron (Broshears and others, 1996; Runkel and others, 1996b). Here we implement the approach of Broshears and others (1996) in a generic manner to fix the dissolved fraction of two coupled components. This approach is empirical in nature, in that specific oxidation/reduction reactions are not modeled. Despite this empirical basis, the fixed fraction approach provides a simple mechanism whereby the proper ratio of two coupled components is maintained. In addition, the parameter governing oxidation/reduction (θ^{target}; eqs. 40 and 41, below) is based on commonly available field data (for example, measurements of dissolved ferrous and ferric iron).

To begin, consider two aqueous components that exchange solute mass as the result of oxidation/reduction reactions. The total dissolved concentration, C^{tot}, is the simple sum of all dissolved species for the two components:

$$C^{tot} = C^1 + C^2 \tag{38}$$

where superscripts "1" and "2" denote the first and second components, respectively. The process of oxidation/reduction is then considered by specifying a single parameter, θ^{target}, to dictate the fraction of the total dissolved concentration that is associated with the first component.

A schematic diagram of the fixed fraction approach is given in figure 9. To implement the approach, an additional loop is placed around the sequential iteration procedure described in Section 2.2.3 (fig. 2). Within each time step, this loop provides an iterative procedure for achieving θ^{target}. As each iteration begins, the sequential iteration procedure is used to determine component concentrations as a function of transport and chemistry. After convergence of the sequential iteration technique, the total dissolved concentration is computed, as well as the fraction of the total dissolved concentration associated with the first component (θ), for each stream segment. This fraction is then compared with that from the previous iteration to test for convergence of the outer loop. The convergence test is given by:

$$|\theta^{m+1} - \theta^m| < 0.005 \tag{39}$$

where m denotes the iteration. If equation 39 holds for all segments, the outer loop has converged and a new time step is initiated. If the solution has not converged, total component concentrations for the two components, T^1 and T^2, are fixed using:

$$T^1 = \theta^{target}C^{tot} + P^1 + S^1 \tag{40}$$

$$T^2 = (1 - \theta^{target})C^{tot} + P^2 + S^2 \tag{41}$$

where P and S denote total precipitate and total sorbed concentrations as defined in Section 2.2.2. External source sink terms for the two components are computed by comparing the quantities given by equations 40 and 41 with values from the previous iteration [for example, $s^1_{ext} = (T^{1,m+1} - T^{1,m})/\Delta t$]. A new iteration of the oxidation/reduction loop then begins.

Oxidation/Reduction

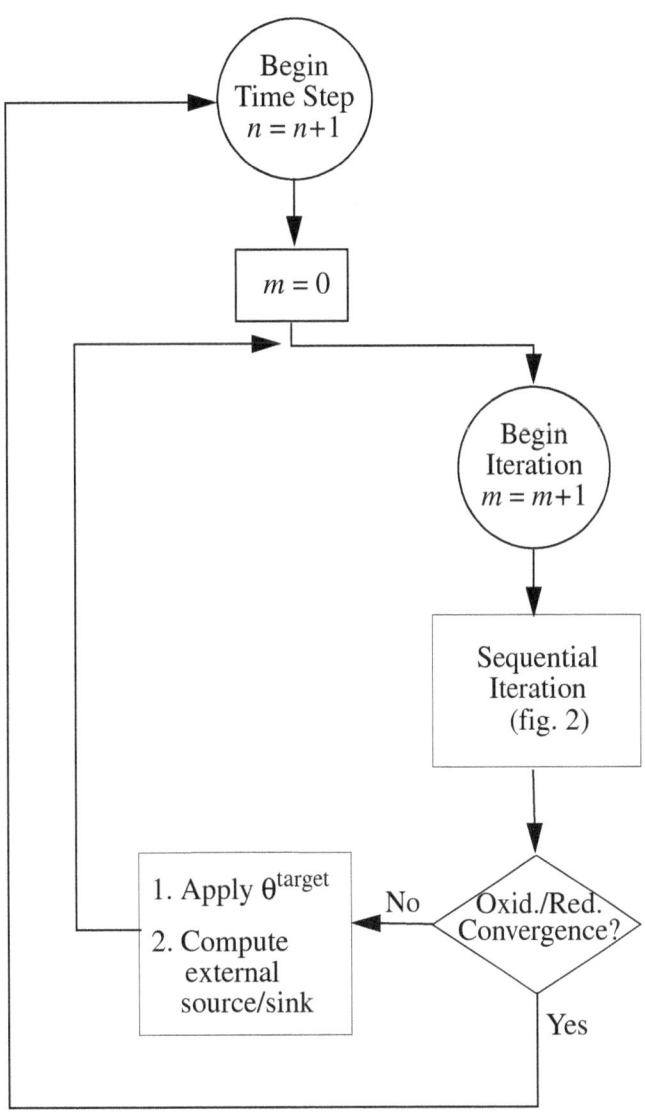

Figure 9. Iterative scheme for oxidation/reduction.

2.3.5   Transient Storage

As given by equation 8, the transport operator describes the physical processes of transient storage, advection, dispersion, and lateral inflow. By including transient storage, equation 8 introduces the storage zone concentration, represented generally as $\hat{C}_S$. Due to the introduction of the storage zone concentration, additional equations and solution techniques are required. The reactive transport problem that includes transient storage is illustrated schematically in figure 10. The top part of the figure represents the main channel, while the bottom part represents the storage zone. Several assumptions inherent to the transient storage formulation are shown in the figure. First, mass in the water column of the main channel is subject to transport processes as noted by the transport operator, L. Mass in the storage zone, meanwhile, is not affected by transport processes. Second, all of the chemical processes described for the main channel also take place in the storage zone. Finally, dissolved and mobile solid concentrations exchange with the storage zone through first-order mass transfer. This is represented by the dotted lines connecting C, P_w, and S_w to their storage zone counterparts, C_s, P_{sw}, and S_{sw}.

This last point is an important assumption for the model presented here. Due to the empirical nature of the transient storage approach, the storage zone represents both open water (eddies and zones of stagnant water) and water within the hyporheic zone (flow through porous areas within the streambed). This lumped system is not problematic for conservative solute transport models, as the solutes of interest are often dissolved-phase tracers that are transported through the hyporheic zone with the water. For the case of reactive solute transport, the modeled solutes are composed of both dissolved and mobile solid phases. If the storage zone is primarily open water, exchange of both dissolved and solid phases between the main channel and the storage zone is a plausible assumption. If the storage zone consists of water in the hyporheic zone, exchange of solid phases between the two regimes is a questionable assumption, as solid mass entering the storage zone may not reenter the main channel due to formation of bonds between the solid and the porous media that comprises the hyporheic zone.

The lumped nature of the storage zone therefore presents a problem for model development. If it is known *a priori* that the storage zone consists of open water, the correct model formulation involves a transient storage mechanism that affects both dissolved and solid phases. Conversely, if the storage zone is primarily the hyporheic zone, the storage mechanism may be set up to affect the dissolved phase only. In reality, storage zones are usually some combination of open water and the hyporheic zone. In the development that follows, the open-water scenario is assumed; dissolved and solid phases are subject to transient storage. This is a logical choice: it correctly models the open water portions of the storage zone, and is also applicable to the hyporheic zone, if the solid particles are small relative to the pore space, such that solid-phase transport occurs.

To implement the transient storage approach, mass-balance equations are developed to define the storage zone concentrations. The total component concentration in the storage zone, T_s, is given by:

$$T_s = C_s + P_s + S_s \tag{42}$$

$$P_s = P_{sw} + P_{sb} \tag{43}$$

$$S_s = S_{sw} + S_{sb} \tag{44}$$

where C_s, P_s, and S_s are the total dissolved, precipitated, and sorbed concentrations, respectively, and P_{sw}, P_{sb}, S_{sw}, and S_{sb} are the mobile (waterborne) precipitate, immobile (bed) precipitate, mobile sorbed, and immobile sorbed concentrations, respectively. As with the main channel, conservation of mass applies to the five phases that make up the total concentration. Development of the five mass-balance equations is similar to that given for the main channel, with the exception that the "mobile" phases of the storage zone (C_s, P_{sw}, S_{sw}) are not affected by advection, dispersion, or lateral inflow. The five equations are summed yielding a governing equation for T_s:

$$\frac{dT_s}{dt} = \alpha \frac{A}{A_s}(T - P_b - S_b - T_s + P_{sb} + S_{sb}) + s_{sext} \tag{45}$$

where A_S is the storage zone cross-sectional area [L^2] and s_{sext} is a source/sink term representing external gains and losses for the storage zone. The main channel concentrations T, P_b, and S_b are given by equations 10–12 (Section 2.2.3). The remaining concentrations are defined by:

$$\frac{dP_{sb}}{dt} = \frac{v_2}{d_2}(P_s - P_{sb}) - f_{sb} \tag{46}$$

$$\frac{dS_{sb}}{dt} = \frac{v_2}{d_2}(S_s - S_{sb}) - g_{sb} \tag{47}$$

Conceptual Surface-Water System with Transient Storage

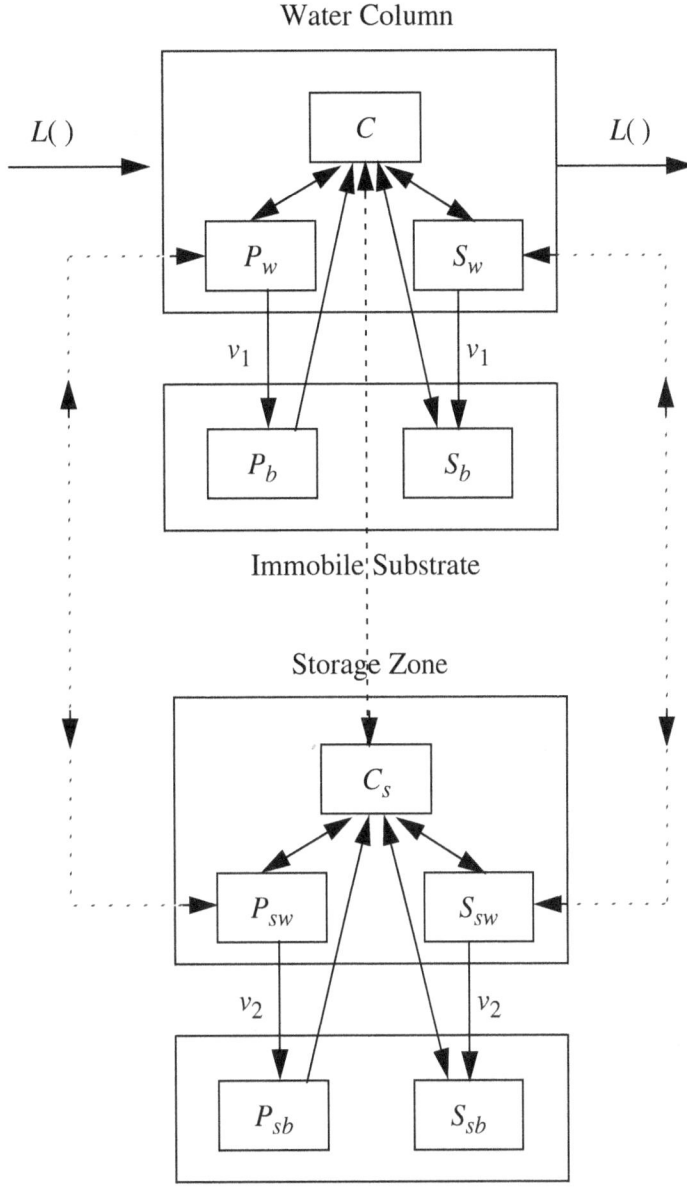

Figure 10.   Conceptual surface-water system, including transient storage, used to develop the governing differential equations. The total component concentration is as defined in figure 1. The total storage-zone component concentration consists of dissolved (C_s), mobile precipitate (P_{sw}), immobile precipitate (P_{sb}), mobile sorbed (S_{sw}), and immobile sorbed (S_{sb}) phases.

where

f_{sb} source/sink for dissolution from the storage zone immobile substrate [moles per liter T^{-1}];

g_{sb} source/sink for sorption/desorption from the storage zone immobile substrate [moles per liter T^{-1}];

v_2 storage zone settling velocity [LT^{-1}]; and

d_2 effective storage zone settling depth [L].

The transient storage equations are easily incorporated into the solution technique presented in Section 2.2.3. First, the initialization phase (Step 1) involves the additional task of estimating total component concentrations for the storage zone ($\tilde{T}_s^{n+1}$). These concentrations are used as input to the equilibrium submodel that determines the dissolved, precipitated, and sorbed concentrations for the storage zone (Step 2). The results from the equilibrium submodel are then used to estimate the source/sink terms for the storage zone in a manner analogous to that used for the main channel. Equilibrium calculations are now required for both the main channel and the storage zone in each stream segment. In Step 3, equations for the immobile phases in the main channel and the storage zone are solved for the dependent variables (eqs. 11, 12, 46, and 47 are solved for P_b, S_b, P_{sb}, and S_{sb}). Equation 10 is then solved for the total component concentration in the main channel, T, using the Crank-Nicolson method and the decoupling procedure described by Runkel and Chapra (1993, 1994). Finally, equation 45 is solved for the total storage zone component concentration, T_s.

2.3.6 Settling of Solid Phases

Equations 11 and 12 model the settling of precipitated and sorbed mass using the settling rate, v_1/d_1. The effective settling depth, d_1 is determined from the depth-area power function:

$$d_1 = aA^b \tag{48}$$

where a is the coefficient of the power function and b is the exponent. Two special cases of equation 48 are of note: (1) for a rectangular channel, $b=1$ and $a=1/w$, where w is channel width; and (2) depth may be specified directly (using coefficient a) by setting $b=0$ ($d_1=aA^0=a$).

2.4 Numerical Solution

The governing equations described above (eqs. 10–12 and 45–47) include a number of spatial and temporal derivatives that must be approximated using numerical methods. The finite-difference techniques used to approximate the derivatives are generally consistent with those presented for the OTIS solute transport model (Runkel, 1998) and will not be repeated here. Additional nomenclature is presented herein, however, to further define the conceptual system used to implement the numerical approach.

2.4.1 The Conceptual System — Segmentation

To implement a numerical solution scheme, the physical system must first be defined. Figure 11 depicts an idealized system in which the stream is subdivided into N discrete segments. Each of these segments represents a control volume within which mass is conserved. Equations 10–12 and 45–47 therefore apply to each segment in the modeled system. Under this segmentation scheme, the subscripts i, $i-1$, and $i+1$ denote concentrations and parameters at the center of three arbitrary segments, while the subscripts $(i-1,i)$ and $(i,i+1)$ define values at the segment interfaces. The length of each segment is denoted by Δx.

Segmentation Scheme

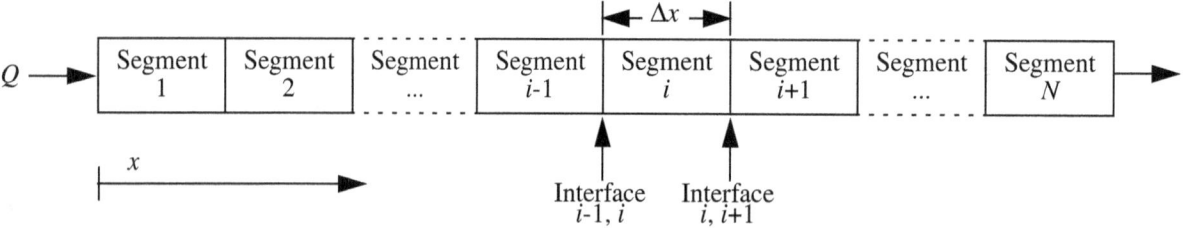

Figure 11. Segmentation scheme used to implement the numerical solution. The conceptual system is subdivided into a number of discrete segments (control volumes).

2.4.2 Boundary Conditions

Equation 10 is solved using finite-difference approximations of the spatial derivatives ($\partial \hat{C}/\partial x$, eq. 8). As with the OTIS solute transport model (Runkel, 1998), a central differencing scheme is used. Using central differences, the concentration in segment i is dependent on the concentrations in the neighboring segments (i-1, i+1). Boundary conditions are therefore needed for the first and last segments of the modeled system (segment i-1 is undefined for segment 1, whereas segment i+1 is undefined for segment N; fig. 11).

Upstream Boundary Condition (Segment 1). The upstream boundary condition is defined in terms of a fixed concentration at the upstream boundary ($\hat{C}_{bc}$, fig. 12). This boundary condition therefore represents the solute concentration entering the upstream end of the modeled system. As such, observed concentration data may be used to satisfy the upstream boundary condition, as discussed in Section 3.3.4 and Section 4.

Segment 1 — Upstream Boundary Condition

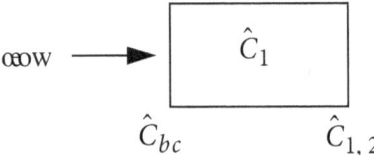

Figure 12. Upstream boundary condition defined in terms of a fixed concentration, $\hat{C}_{bc}$.

Downstream Boundary Condition (Segment N). Direct application of central differencing to the last segment of the modeled system results in the introduction of a concentration term for segment N+1 ($\hat{C}_{N+1}$, fig. 13), a segment that does not exist. Because segment N+1 is outside the modeled system, concentration $\hat{C}_{N+1}$ must be eliminated from the difference equations. This task is accomplished by defining a dispersive flux at the downstream boundary (interface of segment N, N+1; fig. 13):

$$\left(D\frac{\partial \hat{C}}{\partial x} \right)\Bigg|_{N,N+1} = DSBOUND \tag{49}$$

where $DSBOUND$ is a user-supplied value for the dispersive flux. In most applications, $DSBOUND$ is set equal to zero, such that the concentration gradient (change in concentration with respect to space, $\partial \hat{C}/\partial x$) at the downstream boundary is equal to zero. Because non-zero concentration gradients are likely to exist in many systems, the downstream boundary of the modeled system should be placed well downstream of the spatial locations of interest, such that any error associated with the specification of $DSBOUND$ is minimized. User specification of $DSBOUND$ is discussed in Section 3.3.4; placement of the downstream boundary and the potential error is discussed in Section 4.1.

Segment N — Downstream Boundary Condition

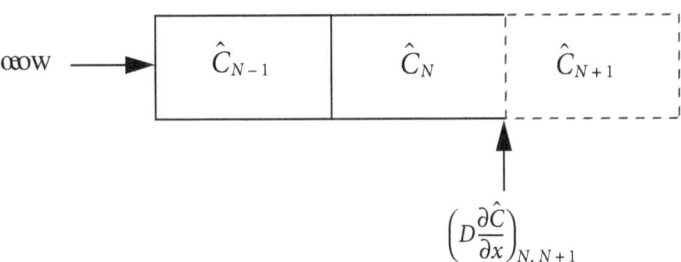

Figure 13. Downstream boundary condition defined in terms of a fixed dispersive flux.

2.4.3   Initial Conditions

In addition to the boundary conditions described in Section 2.4.2, solution of the governing equations requires the specification of initial conditions that define the state of the modeled system at time zero. Initial concentrations must therefore be set in each segment of the stream network (fig. 11). This initialization is accomplished using the initial upstream boundary concentration ($\hat{C}_{bc}$, fig. 12) and a steady-state solution of the governing transport equations (eqs. 10 and 45) in which conservative transport is assumed (Runkel, 1998; physical transport only — in the absence of reactions, $P_b=S_b=S_{ext}=P_{sb}=S_{sb}=S_{sext}=0$). Because the steady-state solution neglects the effects of reaction, the assigned initial concentrations do not reflect the actual effect of the initial upstream boundary condition on the modeled system. The model must therefore be run for a period of time with the initial boundary condition in place prior to reaching a quasi-steady state in which segment concentrations reflect the effects of both physical transport and reaction. This use of model "spin-up" (Thornton and Rosenbloom, 2005) to reach quasi-steady state is illustrated in Section 4.4.

3   User's Guide

This section provides information on the use of the OTEQ solute transport model. OTEQ solves the governing equations described in Section 2.2 based on user-specified model parameters, flow information, and system configuration. These user-specified items are described in Sections 3.1–3.5.

3.1   Conceptual System, Revisited

Before giving a detailed description of the model's input requirements, it is useful to define some of the program variables in terms of the conceptual system. Figure 14 depicts the modeled system as a series of reaches. For our purposes, a reach is defined as a continuous distance along which the physical model parameters remain constant. A reach, for example, will have a spatially constant dispersion coefficient, lateral inflow concentration, and lateral inflow rate. The number of reaches defined for a given system reflects both its inherent variability and the availability of data. A spatially uniform stream may be modeled using a single reach, whereas a stream with a well-characterized variation in channel properties may be simulated using several reaches. The number of reaches in the modeled system is specified by the NREACH input parameter.

Conceptual System — Reaches and Segments

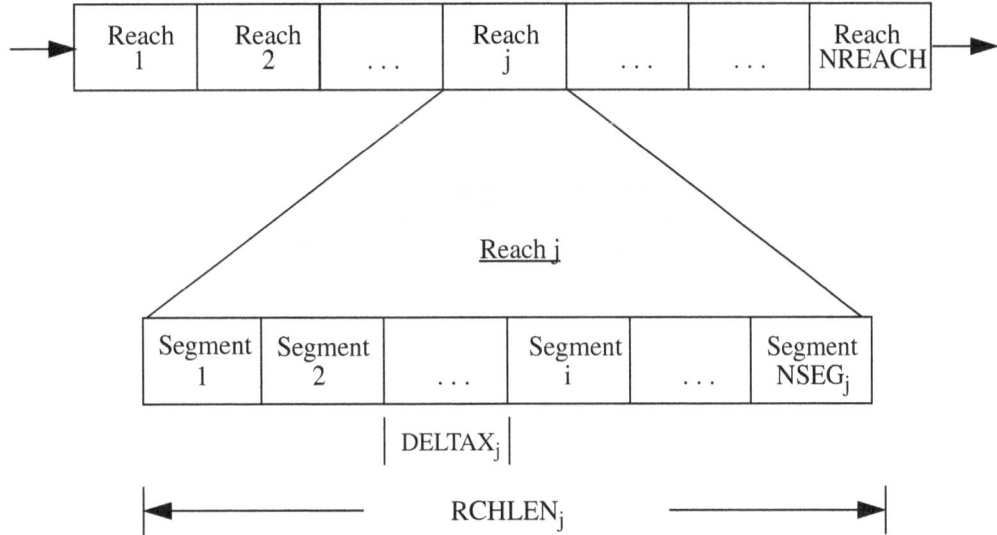

Figure 14.   Conceptual system that includes one or more reaches. Each reach is subdivided into a number of computational elements or segments.

Each reach is subdivided into a number of computational elements or segments. Each segment represents a control volume over which the governing mass-balance equations apply. For a given reach, there are NSEG segments of length DELTAX. Note that DELTAX is determined from the reach length, RCHLEN, and the number of segments:

$$DELTAX = \frac{RCHLEN}{NSEG} \tag{50}$$

Additional program variables are depicted in figure 15, where the first reach in the stream network is shown. Because this reach begins at the upstream boundary of the system, an incoming flow rate, QSTART, and a solute boundary value, USBC, are defined. USBC is the upstream boundary condition, which is discussed in Sections 2.4.2 ($\hat{C}_{bc}$, fig. 12) and 3.3.4 (record type 28, table 16). The program variable denoting the starting stream distance, XSTART, also applies at the upstream end of reach 1. Note that these three variables apply only to the first reach of the modeled system.

Conceptual System — Reach 1

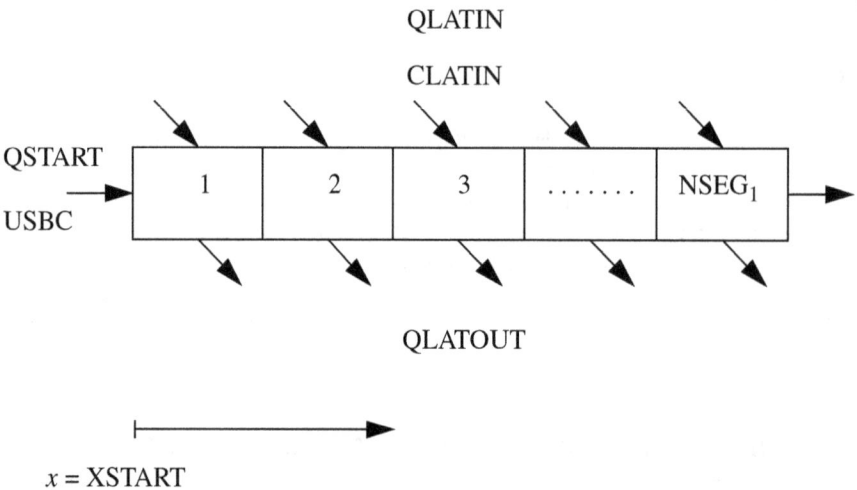

Figure 15.   The first reach in the conceptual system and the required input variables.

The remaining program variables shown in the figure, QLATIN, CLATIN and QLATOUT, are typically specified for each reach in the modeled system (an exception is the case of unsteady flow; see Section 3.3.5). The lateral inflow rate, QLATIN, represents flow entering the channel through surface inflows, overland flow, interflow, and ground-water discharge. This additional flow carries a solute concentration, CLATIN. The final variable, QLATOUT, is a lateral outflow term representing loss of water from the main channel due to ground-water recharge or surface-water diversions.[4] Outflowing water carries a solute concentration that is equal to the simulated main channel concentration. Both QLATIN and QLATOUT are specified on a per unit length basis $[L^3 T^{-1} L^{-1}]$.

3.2   Input/Output Structure

This section describes the input and output files associated with OTEQ. A brief description of each file is presented here; more detailed descriptions are provided in Sections 3.3–3.5.

The input/output structure of OTEQ is depicted in figure 16. As shown in Section 3.3.3, the control file (control.inp) contains user-specified filenames for several input and output files. The parameter file (Section 3.3.4) sets simulation options, boundary conditions, and model parameters that remain constant throughout the simulation. In contrast, the flow file (Section 3.3.5) contains model parameters that can potentially vary during the simulation (for example, volumetric flow rate and main channel cross-sectional area). The control, parameter, and flow files are very similar in form to the corresponding files for the OTIS solute transport model (Runkel, 1998). The MINTEQ input file contains the chemical parameters associated with the equilibrium submodel (Section 3.3.6). Finally, the MINTEQ database files are used to define the reactions considered within the equilibrium submodel (Section 3.3.7).

The filenames for the control file (control.inp) and MINTEQ database files (thermo.dbs, type6.dbs, comp.dbs, analyt.dbs, error.dbs) are set within the software and cannot be changed. The filenames of the parameter, flow, and MINTEQ input files are specified by the user in the control file.

---

[4]Loss of water due to evaporation is not explicitly considered within OTEQ. Implicit consideration of evaporation is possible, however (Keefe and others, 2004); contact the author of this report for additional details.

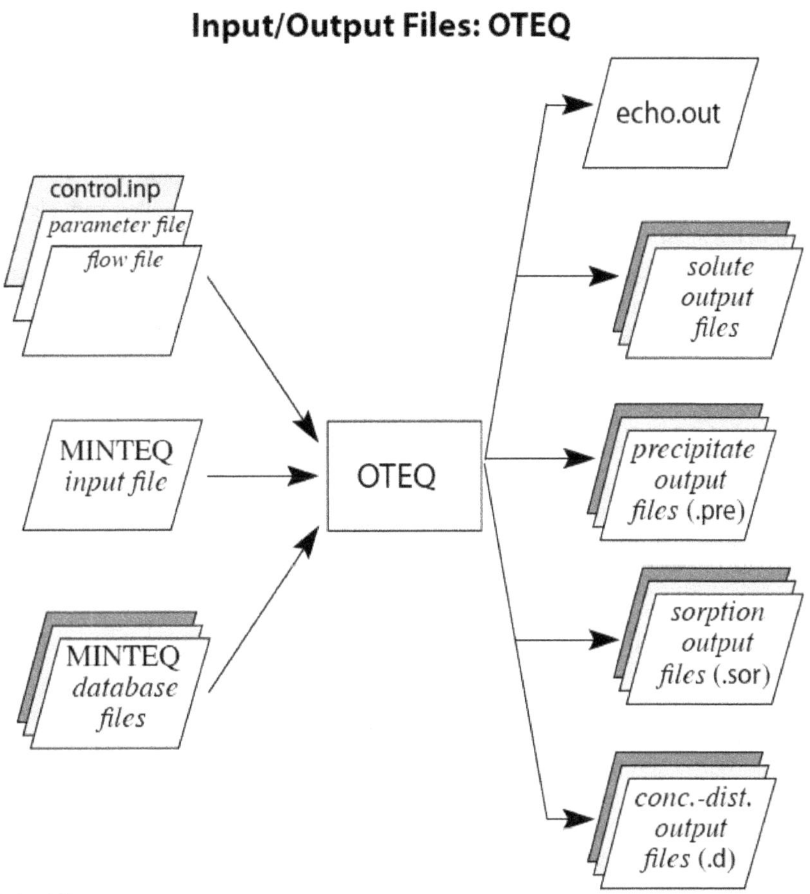

Input/Output Files: OTEQ

Figure 16. OTEQ Input/Output files.

Also shown in figure 16 are the output files created by OTEQ. Upon completion of a model run, the file **echo.out** contains an "echo" of the user-specified simulation options and model parameters, a summary of iteration information at each time step, and any error messages generated during model execution. In addition to **echo.out**, the model creates one solute output file for each solute. If precipitation is being modeled, a precipitate output file is created for each solute involved in precipitation reactions. Similarly, when sorption is modeled, one sorption output file is also created for each solute involved in sorption reactions. As described in Section 3.5.1, the solute, precipitate, and sorption output files contain a time-series of solute concentrations at the user-specified print locations (Section 3.3.4).

Concentration-distance output files are optional output files that are created at the user's request. These files contain a spatial profile of the simulated solute concentrations at the completion of the model run. Additional details on the concentration-distance output files are provided in Section 3.5.2.

The filenames for the solute output files are specified by the user in **control.inp**. Filenames for the precipitate and sorption output files are created by adding a ".**pre**" and ".**sor**" extension to the corresponding solute output filename. Filenames for the concentration-distance output files are created by adding a ".**d**" extension to the solute, precipitate, and sorption output filenames.

3.3   Input Format

As described in Section 3.2, several input files must be assembled prior to model execution. In the sections that follow, each input file is described in terms of a set of record types. Within each record type, one or more fields are used to specify various input parameters. In general, record types refer to lines (rows) in the input file, and fields correspond to specific columns within each record. In most cases, the specific columns for a given field are specified in the table that describes each record type. Some exceptions to the column requirements are noted in the text, when a given record type is in a free format and placement of input in specific columns is not required. In addition to column specifications, each table specifies the field formats. A given field may be double precision (D), integer (I), or character (C). Double precision fields require entry of a numeric value that includes a decimal point; integer fields require the entry of a numeric value without a decimal point. Alphanumeric values may be placed in character fields. Example input files are presented in Section 4.

3.3.1   Units

In Section 2.2 the governing equations and model parameters are described using the fundamental units of length [L] and time [T]. Within OTEQ, specific units are assigned to the various input parameters. The model user may select the appropriate units subject to the following:

- **Length Units [L].** Any unit of length may be used when specifying model input variables. The only requirement is that the length unit must be consistent for all model parameters and system configuration variables. The chosen length unit is denoted by an L in the input file descriptions that follow.

- **Time Units.** All model parameters that require the specification of a time unit (flow rates and dispersion coefficients, for example) are defined in terms of seconds. Note, however, that simulation control variables such as the simulation start time (TSTART) are specified in terms of hours.

- **Concentration Units.** Concentration units of moles per liter are required for all model quantities involving concentration (boundary conditions, lateral inflow concentrations, etc.).

3.3.2   Internal Comments

Model users may document their work by placing comments within any of the input files described below. All lines with a pound sign (#) placed in column number 1 will be treated as comments by the model. This feature is illustrated in the example input files described in Section 4.

3.3.3   The Control File

The control file, control.inp, is used to specify the filenames for the various input and output files. The format of the control file is described here; an example control file is presented in Section 4.1. The format of control.inp is shown in table 1, where the control file consists of four record types. Record type 1 specifies the filename for the parameter file, record type 2 specifies the name of the MINTEQ input file, record type 3 specifies the name of the flow file, and record type 4 specifies the names of the solute output files. Although table 1 lists four record types, the control file will contain more than four records, as record type 4 is repeated for each solute (output files are created for each solute).

Table 1.   The OTEQ control file.

[C, Character]

Record type	Input variable	Format	Column	Description
1	FILE	C	1–20	Filename for the *Parameter File*
2	FILE	C	1–20	Filename for the MINTEQ *Input File*
3	FILE	C	1–20	Filename for the *Flow File*
4[1]	FILES	C	1–20	Filenames for the *Solute Output Files*

[1]Record type 4 is used once for each solute modeled (NSOLUTE times).

3.3.4 The Parameter File

The parameter file specifies print options, boundary conditions, and the model parameters that remain constant throughout the run. The format of the parameter file is given in tables 2–16. The parameter file is created using the 28 record types discussed below.

Record Type 1 — Simulation Title

The first record in the parameter file is a simulation title of up to 80 characters. This title is printed as part of the echo.out.

Record Type 2 — Print Step

The print step specifies the time interval at which results are printed to the solute output files. If, for example, results are needed every 15 minutes, a value of 0.25 hour is entered for the print step. The actual print step and the requested print step may differ if the requested step is not an even multiple of the integration time step (record type 3). In this case, the model sets the print step equal to the nearest multiple of the integration time step.

Record Types 3–5 — Time Parameters

The next step in constructing the parameter file is to enter the appropriate time parameters. Record type 3 is used to set the TSTEP input variable. TSTEP is the integration time step (Δt) used within the numerical solution. Multiple model runs should be performed to determine a time step that yields an accurate solution.

The remaining time parameters, TSTART and TFINAL, are set using record types 4 and 5. The input variable TSTART denotes the simulation start time, and TFINAL specifies the simulation end time. As shown in table 2, values for TSTEP, TSTART, and TFINAL are specified in hours.

Record Type 6 — Distance at the Upstream Boundary

Record type 6 specifies XSTART, the distance at the upstream boundary of the modeled system. During model execution, XSTART is used to determine the distances at various locations downstream. As the upstream boundary is at the beginning of the modeled area, XSTART is commonly set to 0.0.

Record Type 7 — Downstream Boundary Condition

The downstream boundary condition, DSBOUND, is set using record type 7. In many modeling applications, the flux represented by the downstream boundary condition is set to zero. As discussed in Section 2.4.2, setting DSBOUND to zero implies that the concentration gradient at the downstream boundary is equal to zero. Due to this assumption, the length of the modeled system should be such that the location of the downstream boundary is sufficiently downstream from the nearest location of interest. Additional details on the downstream boundary condition are provided in Sections 2.4.2 and 4.1.

Record Type 8 — Relative Error Tolerance

The relative error tolerance, TOL, is specified using record type 8. The relative error tolerance is used to check for convergence of the sequential iteration procedure (Section 2.2.3, eq. 14).

Record Type 9 — Chemistry Option

As described in Section 2.2.3, the equilibrium submodel is called for each segment in the modeled system to determine the chemical composition of water in the main channel. Section 2.3.5 describes the corresponding procedure for determining the composition of the transient storage zone. OTEQ is configured such that the user has the option of requesting these additional computations for the storage zone. This option is implemented using the chemistry option, ICHEM. If ICHEM is set to 1, equilibrium computations will be restricted to the main channel and chemical reactions within the transient storage zone will not be modeled. If ICHEM is set to 2, equilibrium computations will be conducted for both the main channel and the transient storage zone. It is important to note that the physical process of transient storage is modeled regardless of the value of ICHEM (provided ALPHA > 0.0, record type 11).

Record Type 10 — The Number of Reaches

As discussed in Section 3.1, the modeled system is divided into a number of reaches, NREACH. This parameter is set by record type 10. For each reach, spatially constant parameters are specified using record type 11.

Table 2. The parameter file — record types 1–10.

[C, Character; D, Double precision; I, Integer; L, chosen length unit (Section 3.3.1)]

Record type	Input variable	Format	Column	Units	Description
1	TITLE	C	1–80	—	Simulation title
2	PSTEP	D	1–13	hours	Print step
3	TSTEP	D	1–13	hours	Integration time step
4	TSTART	D	1–13	hour	Simulation starting time
5	TFINAL	D	1–13	hour	Simulation ending time
6	XSTART	D	1–13	L	Distance at the upstream boundary
7	DSBOUND	D	1–13	(L per second)(moles per liter)	Downstream boundary condition
8	TOL	D	1–13	—	Relative error tolerance
9	ICHEM	I	1–5	—	Chemistry option (1 or 2)
10	NREACH	I	1–5	—	Number of reaches

Record Type 11 — Reach Specific Parameters

Record type 11 sets the parameters for each reach (table 3). Unlike record types 1–10, record type 11 contains more than one input variable. The first field in the record indicates the number of segments located within the reach, NSEG, and a second field specifies the length of the reach, RCHLEN. These two parameters define the length of each segment, DELTAX, as described in Section 3.1.

The dispersion coefficient (DISP), storage zone cross-sectional area (AREA2), and the storage zone exchange coefficient (ALPHA), are specified in fields 3–5. The transient storage mechanism may be turned off by setting ALPHA to 0.0, as shown in Application 1 (Section 4.1). To avoid division by zero, AREA2 must be set to a non-zero value. This value will not affect simulation results, however, provided ALPHA is 0.0.

The coefficient in the depth-area power function (Section 2.3.6, eq. 48), A1, and the exponent in the depth-area power function, B1, complete record type 11. Because record type 11 is used for reach-specific parameters, it is used NREACH times, once for each reach in the stream network.

Table 3. The parameter file — record type 11.

[D, Double precision; I, Integer; L, chosen length unit (Section 3.3.1); Record type 11 is used once for each reach (NREACH times)]

Input variable	Format	Column	Units	Description
NSEG	I	1–5	—	Number of segments in reach
RCHLEN	D	6–15	L	Reach length
DISP	D	16–25	L^2 per second	Dispersion coefficient
AREA2	D	26–35	L^2	Storage zone cross-sectional area
ALPHA	D	36–45	per second	Storage zone exchange coefficient
A1	D	46–55	—	Coefficient in depth-area power function
B1	D	56–65	—	Exponent in depth-area power function

Record Types 12 and 13 — Solute Information

Governing equations for the reactive transport model are derived in Section 2.2. These equations describe the transport of the chemical components within the modeled system. With the exception of the sorption components (Section 2.3.3, eqs. 24 and 25), all components within the equilibrium submodel are subject to transport as defined by the governing equations. Within OTEQ, these transported components are known as solutes.

The number of solutes, NSOLUTE, is specified in record type 12 (table 4). In accordance with the above discussion, NSOLUTE is equal to the number of components in the MINTEQ input file (Section 3.3.6), less the number of sorption components.

Table 4. The parameter file — record type 12.

[I, Integer]

Input variable	Format	Column	Description
NSOLUTE	I	1–5	Number of solutes

Record type 13 defines each solute (table 5). The first field in the record specifies the MINTEQ component number, ID, associated with a given solute. The ID specified here must also be present in the MINTEQ input file (Section 3.3.6); the correspondence between MINTEQ component numbers (ID) and specific chemical constituents is given by Allison and others (1991). Fields 2–4 contain flags which indicate whether a given solute is involved in precipitation reactions (PFLAG), sorption reactions (SFLAG), and (or) reactions involving an external source/sink (EFLAG). A flag value of 1 indicates that the solute participates in the relevant reactions, whereas a value of 0 indicates that the solute does not participate. PFLAG should be set to 1 for all solutes (components) that make up the precipitate species defined using record type 15. SFLAG should be set to 1 for all solutes that participate in GTLM sorption reactions (see Section 3.4.3). Because record type 13 defines each solute, it is used NSOLUTE times.

Table 5. The parameter file — record type 13.

[I, Integer; Record type 13 is used once for each solute modeled (NSOLUTE times)]

Input variable	Format	Column	Description
ID	I	1–5	MINTEQ component number
PFLAG	I	6–10	Precipitation flag (0 or 1)
SFLAG	I	11–15	Sorption flag (0 or 1)
EFLAG	I	16–20	External source/sink flag (0 or 1)

Record Type 14 — Number of Precipitates

OTEQ allows for the precipitation of one or more precipitate species. The number of precipitate species, NPRECIPS, is specified using record type 14 (table 6). When precipitation is modeled, NPRECIPS is set equal to the number of possible solids defined within the equilibrium submodel (possible solids are defined in the MINTEQ input file; see Sections 3.3.6 and 3.4.3). If precipitation reactions are not modeled, NPRECIPS is set to 0 and record type 15 is not used.

Table 6. The parameter file — record type 14.

[I, Integer]

Input variable	Format	Column	Description
NPRECIPS	I	1–5	Number of precipitate species

Record Type 15 — Precipitate Definition (optional)

Note: Record type 15 is omitted if precipitation is not modeled (NPRECIPS=0, record type 14).

Precipitate species are defined using record type 15 (table 7). The first field in record type 15 specifies the MINTEQ identification number of the precipitate species, IDPRECIP. IDPRECIP must be the identification number of a species that is defined as a possible solid within the MINTEQ input file (Sections 3.3.6 and 3.4.3). Fields two and three are used to specify the main channel and storage zone settling velocities, PSETTLE and PSETTLE2, respectively (Section 2.2.3, eqs. 11 and 12; Section 2.3.5, eqs. 46 and 47). Record type 15 is specified for each precipitate species (it is used NPRECIPS times).

Table 7. The parameter file — record type 15.

[D, Double precision; I, Integer; L, chosen length unit (Section 3.3.1); Record type 15 is used once for each precipitate (NPRECIPS times)]

Input variable	Format	Column	Units	Description
IDPRECIP	I	1–8	—	MINTEQ identification number of precipitate
PSETTLE	D	9–18	L per second	Main channel settling velocity
PSETTLE2	D	19–28	L per second	Storage zone settling velocity

Record Type 16 — Sorption Option

As described in Section 2.3.3, OTEQ may model sorption onto static and (or) dynamic surfaces. The type of sorption modeled is controlled by the sorption option, ISORB (record type 16, table 8). If sorption is not modeled, ISORB is set to 0. A value of 1 indicates sorption to a static surface, whereas a value of 2 indicates sorption to a dynamic surface. Concurrent sorption onto static and dynamic surfaces is requested by setting ISORB to 3. Sorption onto multiple static and (or) multiple dynamic surfaces is not currently possible.

Table 8.   The parameter file — record type 16.

[I, Integer]

Input variable	Format	Column	Description
ISORB	I	1–5	Sorption option (0, 1, 2, or 3)

Record Types 17–20 — Sorption Definition: Static Surface (optional)

Note: Record types 17–20 are omitted if sorption to a static surface is not modeled (ISORB=0 or 2, record type 16).

Record types 17–20 are specified if sorption to a static surface is considered (ISORB=1 or 3). Sorption to a static surface may be in chemical equilibrium or kinetically limited. The kinetic rate coefficient, LAMBS, is set using record type 17 (table 9). For kinetically limited sorption, a kinetic rate coefficient is defined as:

$$LAMBS = \frac{\Gamma}{\Delta t} \tag{51}$$

where Γ is the fraction of the equilibrium quantity allowed to sorb/desorb during a time step ($0.0 < \Gamma < 1.0$; Section 2.3.3, eqs. 30 and 36) and Δt is the integration time step in seconds (TSTEP × 3,600.0). For equilibrium sorption, LAMBS is set to 999.0.

Record type 18 specifies the sorbent solid concentration associated with the static surface, SOLCON (S_C, Section 2.3.3, eqs. 24 and 25). Record type 18 is specified for each reach (it is used NREACH times).

Table 9.   The parameter file — record types 17–18.

[D, Double precision]

Record type	Input variable	Format	Column	Units	Description
17	LAMBS	D	1–13	per second	Kinetic rate coefficient for sorption
18[1]	SOLCON	D	1–13	gram per liter	Sorbent solid concentration

[1]Record type 18 is used once for each reach (NREACH times).

In record type 13, SFLAG is used to indicate which solutes are involved in sorption reactions. To complete the definition of the static surface, record types 19 and 20 are specified for each solute with SFLAG=1. Record type 19 is first used to specify IDSOLID, the MINTEQ component number. Record type 20 is then used to specify the initial immobile sorbed concentration of IDSOLID in the main channel (SORBB) and the storage zone (SORB2B) in reach one. After specifying the initial sorbed concentrations in reach one, record type 20 is repeated for the remaining reaches (record type 20 is used NREACH times). The block of record types 19 and 20 is then specified for the remaining solutes with SFLAG=1. Note that the blocks of record types must be specified in the same order as the MINTEQ component numbers (ID) in record type 13.

Table 10.   The parameter file — record types 19–20.

[D, Double precision; I, Integer]

Record type	Input variable	Format	Column	Units	Description
19[1]	IDSOLID	I	1–5	—	MINTEQ component number of solute involved in sorption
20[1,2]	SORBB	D	1–13	moles per liter	Initial immobile sorbed concentration, main channel
20[1,2]	SORB2B	D	1–13	moles per liter	Initial immobile sorbed concentration, storage zone

[1]The block of record types 19–20 is used for each solute with SFLAG=1.

[2]Record type 20 is used once for each reach (NREACH times).

Record Type 21 — Sorption Definition: Dynamic Surface (optional)

Note: Record type 21 is omitted if sorption to a dynamic surface is not modeled (ISORB=0 or 1, record type 16).

Record type 21 is specified if sorption to a dynamic surface is considered (ISORB=2 or 3). Field one of record type 21 (table 11), IDSORB, specifies the MINTEQ component number of the solute whose precipitate makes up the dynamic surface. Field two specifies MWSORB, the molecular weight of the dynamic surface.

Table 11.   The parameter file — record type 21.

[D, Double precision; I, Integer]

Input variable	Format	Column	Units	Description
IDSORB	I	1–7	—	MINTEQ component number of solute associated with dynamic surface
MWSORB	D	8–19	gram per mole	Molecular weight of sorbent

Record Type 22 — Redox Option

Section 2.3.4 presents an iterative procedure whereby oxidation/reduction is considered. Oxidation/reduction is modeled by setting the redox option, IREDOX, to 1 (record type 22, table 12) and specifying additional parameters using record types 23 and 24. Oxidation/reduction will not be modeled if IREDOX is set to 0.

Table 12.   The parameter file — record type 22.

[I, Integer]

Input variable	Format	Column	Description
IREDOX	I	1–5	Redox option (0 or 1)

Record Types 23 and 24 — Redox Definition (optional)

Note: Record types 23 and 24 are omitted if oxidation/reduction is not modeled (IREDOX=0, record type 22).

Record type 23 (table 13) is used to specify which solutes are involved in oxidation/reduction reactions. Fields one and two specify the MINTEQ component numbers of the two solutes involved in oxidation/reduction, IDRED and IDRED2. Record type 24 specifies the target fraction used in the iterative procedure, THETA (θ^{target}, Section 2.3.4). The target fraction may be spatially variable; record type 24 is therefore specified for each reach (it is used NREACH times). As described in Section 2.3.4, the iterative procedure distributes mass between IDRED and IDRED2 such that the fraction of the total dissolved concentration associated with IDRED is equal to THETA (where 0.0 < THETA < 1.0).

Table 13. The parameter file — record types 23 and 24.

[D, Double precision; I, Integer]

Record type	Input variable	Format	Column	Description
23	IDRED	I	1–5	MINTEQ component number
23	IDRED2	I	6–10	MINTEQ component number
24[1]	THETA	D	1–10	Fraction of total dissolved that is IDRED

[1]Record type 24 is used once for each reach (NREACH times).

Record Types 25 and 26 — Output Specifications

Record types 25 and 26 control the format and type of output files created by OTEQ (table 14). OTEQ provides the ability to output the time-series of solute concentration at a number of fixed locations (solute, precipitate, and sorption output files, Section 3.2) and the spatial distribution of solute concentration at a fixed time (concentration-distance output files, Section 3.2).

Within OTEQ, the fixed locations used for time-series output are known as print locations. Field one of record type 25 specifies the number of print locations, NPRINT. Fields 2–4 set the interpolation option (IOPT), the distance option (DOPT), and the print option (PRTOPT). Record type 26 specifies the distance of each print location, PRTLOC, and is used NPRINT times.

If the interpolation option (IOPT) is set to 1, the concentration at each print location is determined by linear interpolation using the centers of the two segments closest to the print location. If IOPT is set to 0, the concentration at each print location is set equal to the concentration of the nearest upstream segment. For many applications, specification of IOPT has a negligible effect on simulation results. Selection of the interpolation option (IOPT=1) may be advantageous when long segment lengths (DELTAX) are used. Interpolation is not recommended when the print locations fall near reach endpoints and the downstream reach is characterized by physical parameters that differ markedly from those of the current reach.

The distance option, DOPT, is used to request the creation of concentration-distance files. If DOPT is set to 1, concentration-distance files will be created that contain the solute concentrations at the end of the simulation (for time equal to TFINAL, record type 5). If DOPT is set to 0, concentration-distance files are not created. Concentration-distance files are most often requested when a steady-state solution is of interest (see Applications 4 and 5, Sections 4.4–4.5).

The print option, PRTOPT, determines the format of the output files. If the print option is set to 1, solute concentrations are output for the main channel only. Solute concentrations in both the main channel and the storage zone are output if the print option is set to 2.

Table 14.   The parameter file — record types 25–26.

[D, Double precision; I, Integer; L, chosen length unit (Section 3.3.1)]

Record type	Input variable	Format	Column	Units	Description
25	NPRINT	I	1–5	—	Number of print locations
25	IOPT	I	6–10	—	Interpolation option (0 or 1)
25	DOPT	I	11–15	—	Distance option (0 or 1)
25	PRTOPT	I	16–20	—	Print option (1 or 2)
26[1]	PRTLOC	D	1–13	L	Print location

[1]Record type 26 is used for each print location (NPRINT times).

Record Types 27 and 28 — Upstream Boundary Conditions

The final record types in the parameter file specify the time-variable upstream boundary condition. As shown in table 15, the number of boundary conditions (NBOUND) and the boundary condition option (IBOUND) are set using record type 27. The NBOUND upstream boundary conditions are characterized by a starting time and a boundary value for each solute using record type 28. Each field in record type 28 has a free format such that placement of input in specific columns is not required. The time at which a boundary condition goes into effect (USTIME) is specified in field one of record type 28 (table 16). Field two, USBC, denotes the boundary values corresponding to USTIME. The USBC field is repeated horizontally for each solute modeled (the boundary value for solute one is placed to the right of USTIME, the boundary value for solute two is placed to the right of USBC for solute one, etc.). After specifying USTIME and USBC for the first boundary condition, record type 28 is repeated for each subsequent change in the boundary condition (it is used NBOUND times).

Table 15.   The parameter file — record type 27.

[I, Integer]

Input variable	Format	Column	Units	Description
NBOUND	I	1–5	—	Number of boundary conditions
IBOUND	I	6–10	—	Boundary condition option (1, 2, or 3)

Table 16. The parameter file — record type 28, upstream boundary conditions.

[D, Double precision; L, chosen length unit (Section 3.3.1); Record type 28 is used once for each boundary condition (NBOUND times)]

Input variable	Format	Units[1]	Description
USTIME	D	hour	Time boundary condition begins
USBC[2]	D	moles per liter; (moles per liter)(L^3 per second)	Upstream boundary value

[1]Units for USBC are dependent on the value of IBOUND in record type 27 (fig. 17).

[2]USBC repeats horizontally (the value for solute one is placed to the right of USTIME, the value for solute two is placed to the right of USBC for solute one, etc.).

Interpretation of USBC is dependent on the type of upstream boundary condition specified. Three types of boundary conditions may be specified using the boundary condition option, IBOUND, as shown in figure 17. A step concentration profile is imposed at the upstream boundary for IBOUND equal to 1. Under this option, USBC corresponds to the upstream boundary concentration, C_{bc}, described in Section 2.4.2. With a step concentration profile, C_{bc} is initially set to the first boundary value ($USBC_j$) and subsequently updated (to $USBC_{j+1}$, $USBC_{j+2}$,…) at the appropriate times ($USTIME_{j+1}$, $USTIME_{j+2}$,…).

Upstream Boundary Condition Options

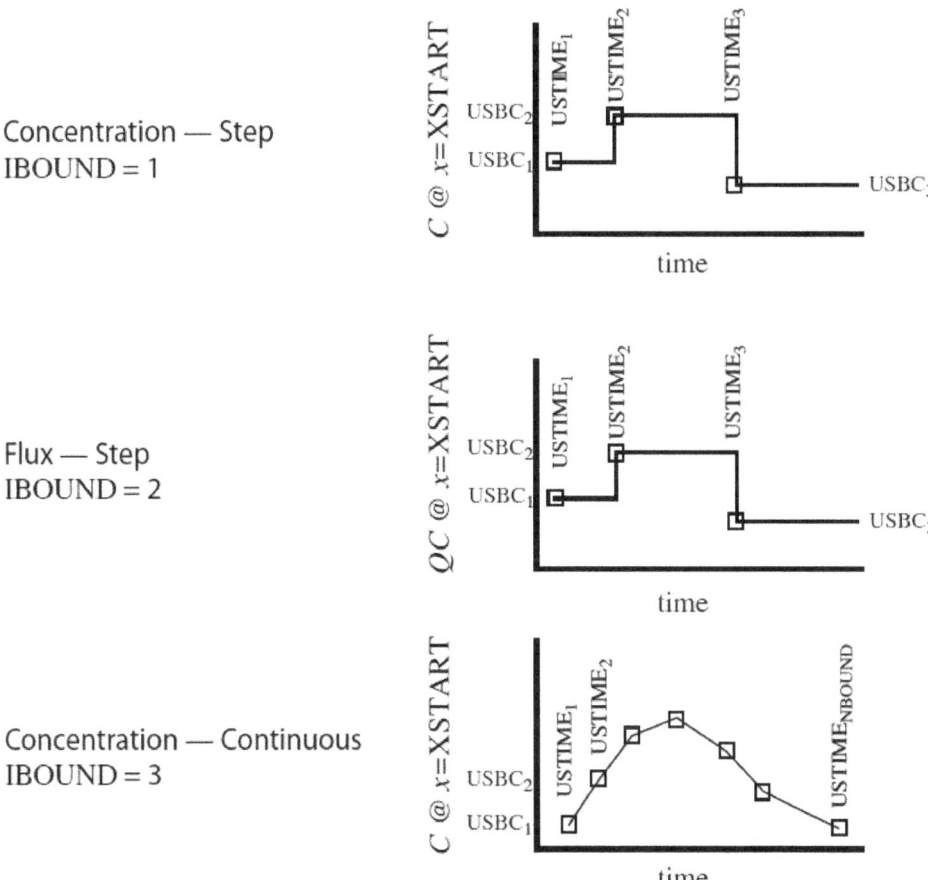

Figure 17. Upstream boundary condition options. The upstream boundary condition may be in terms of a step concentration profile, a step flux profile, or a continuous concentration profile.

A step flux profile is imposed at the upstream boundary for IBOUND equal to 2. Under this option, USBC corresponds to QC_{bc}, where Q is the volumetric flow rate at the upstream boundary (Q=QSTART; figure 15, table 18). With a step flux profile, the first boundary value (USBC$_j$) is divided by Q to obtain an initial value for C_{bc}. The upstream boundary concentration is subsequently updated at the appropriate times (USTIME$_{j+1}$, USTIME$_{j+2}$,…). An example of a step flux boundary condition is presented by Runkel (1998).

A continuous concentration profile is imposed at the upstream boundary for IBOUND equal to 3. Under this option, C_{bc} is updated at each model time step. The value assigned to C_{bc} is determined by linear interpolation using specified values of USBC. To allow for interpolation, the time of the last boundary condition (USTIME$_{NBOUND}$) must be greater than or equal to the simulation end time, TFINAL (record type 5). An example of a continuous concentration boundary condition is presented by Runkel (1998).

3.3.5   The Flow File

The flow file defines the model parameters that can potentially vary in time. These parameters include the volumetric flow rate (Q, QSTART), lateral flow rates (QLATIN, QLATOUT), main channel cross-sectional area (AREA), and lateral inflow solute concentration (CLATIN). Collectively, these parameters are known as the flow variables.

The format of the flow file depends on the nature of the parameters contained therein. If all of the flow variables are constant with respect to time, the flow is steady, whereas if any of the flow variables change in time, the flow regime is considered unsteady.

For steady flow conditions, the flow variables are specified on a reach-by-reach basis. When the flow regime is unsteady, the flow variables are specified using flow locations. The formats of the steady and unsteady flow files are given below.

The Flow File — Steady Flow

When the flow variables are constant with respect to time, the steady flow option is invoked. As shown in the following sections, the specification of a steady flow file is fairly straightforward.

Record Type 1 — Change-in-Flow Indicator. The first record type in the flow file is the change-in-flow indicator, QSTEP. This record type is used to define the flow regime. As shown in table 17, QSTEP is the time interval between changes in the flow variables. Here we are concerned with a steady flow regime in which the flow variables are constant. In this case, QSTEP is set to zero.

Table 17.   Steady flow file — record type 1.

[D, Double precision]

Input variable	Format	Column	Units	Description
QSTEP	D	1–13	hour	Change-in-flow indicator (set to zero)

Record Types 2 and 3 — Flow Variables. The remaining record types for a steady flow file are described in tables 18 and 19. Record type 2 sets the volumetric flow rate at the upstream boundary, QSTART. The remaining flow variables are set using record type 3, where each field has a free format such that placement of input in specific columns is not required. As shown in table 19, the first three fields of this record specify the lateral flow rates (QLATIN and QLATOUT) and the cross-sectional area (AREA) for a particular reach. The fourth field, used to indicate the lateral inflow solute concentration (CLATIN), repeats horizontally for each solute modeled. Because the flow variables vary from reach to reach, record type 3 must be used once for each reach in the network.

Table 18.   Steady flow file — record type 2.

[D, Double precision; L, chosen length unit (Section 3.3.1)]

Input variable	Format	Column	Units	Description
QSTART	D	1–13	L^3 per second	Flow rate at the upstream boundary

Table 19. Steady flow file — record type 3.

[D, Double precision; L, chosen length unit (Section 3.3.1); Record type 3 is used once for each reach in the stream network (NREACH times)]

Input variable	Format	Units	Description
QLATIN	D	L^3 second^{-1} L^{-1}	Lateral inflow rate for reach j
QLATOUT	D	L^3 second^{-1} L^{-1}	Lateral outflow rate for reach j
AREA	D	L^2	Main channel area for reach j
CLATIN[1]	D	moles per liter	Lateral inflow solute concentration

[1]CLATIN repeats horizontally (the lateral inflow concentration for solute one is placed to the right of AREA, the concentration for solute two is placed to the right of CLATIN for solute one, etc.).

The Flow File — Unsteady Flow

With the allowance for time-variable flow parameters comes an increase in model complexity. As a result, the format of the unsteady flow file differs markedly from that presented for the steady case. In addition, the unsteady flow file is much larger than its steady counterpart, as several record types must be repeated for each change in the flow variables.

The format used here is compatible with standard routing models such as DR$_3$M (Alley and Smith, 1982) and DAFLOW (Jobson, 1989). As discussed below, record types 2–8 allow for the specification of time-varying parameters at various locations along the stream channel. These record types may be created by reformatting output from the selected routing model.

Record Type 1 — Change-in-Flow Indicator. For an unsteady flow regime, the flow variables are read from the flow file when changes in the flow values occur. QSTEP therefore defines the times at which the flow variables are updated. If the flow variables change every 15 minutes, for example, QSTEP is set to 0.25 hour and the variables are read at the appropriate times.

Table 20. Unsteady flow file — record type 1.

[D, Double precision]

Input variable	Format	Column	Units	Description
QSTEP	D	1–13	hour	Time interval between changes in flow

Record Types 2 and 3 — Flow Locations. When detailed field measurements are available, it may be appropriate to specify time-varying flow variables at various locations along the stream. These values may be available as output from hydrologic or hydraulic routing models. The format of the flow file allows the user to define a number of flow locations at which the flow variables are specified. These flow locations often correspond to the reaches that are defined in the parameter file. (This correspondence is not required, however.) Using the data provided at the flow locations, the model linearly interpolates flow and area values (Q, AREA) for each segment in the stream network. Lateral inflow rates and concentrations (QLATIN, CLATIN) are also set using the flow locations.

Record type 2 indicates the number of flow locations, NFLOW. Record type 3 specifies the distance of a given flow location. This record is used NFLOW times, once for each flow location. Several requirements must be kept in mind when specifying flow locations. The requirements, listed below, are checked internally by the model:

- The flow locations must be entered in ascending (downstream) order; the second location must be downstream from the first, the third must be downstream from the second, etc.
- The first flow location must be placed at the upstream end of the stream network. This implies that FLOWLOC$_1$ equals XSTART, where XSTART is the starting location specified in the parameter file. Note that the flow specified at this first location is analogous to the QSTART parameter in the steady flow file.
- The last flow location must be placed at or below the downstream boundary.

Table 21.   Unsteady flow file — record types 2 and 3.

[D, Double precision; I, Integer; L, chosen length unit (Section 3.3.1)]

Record type	Input variable	Format	Column	Units	Description
2	NFLOW	I	1–5	—	Number of flow locations
3[1]	FLOWLOC	D	1–13	L	Flow location

[1]Record type 3 is used once for each flow location (NFLOW times).

Record Types 4–7 — Lateral Flows, Flows, and Areas. In contrast to the steady flow file, lateral flows and concentrations (QLATIN, CLATIN) do not necessarily correspond to specific reaches. In the unsteady flow file, these parameters are specified for each flow location. The value specified is used for all segments in between the current flow location and the flow location immediately upstream (from location j-1 to location j). Note that this scheme corresponds to that for the steady flow file if the flow locations are placed at the end of each reach. As stated earlier, the flow and area (Q, AREA) values at the flow locations are used to interpolate values for the segments within the network.

Values for lateral inflow, flow rate, area, and lateral inflow concentration are set for each flow location using record types 4–7, respectively. Record types 4–7 all use a free format such that placement of input in specific columns is not required. As shown in table 22, the input fields in these record types (QLATIN, Q, AREA, CLATIN) repeat horizontally for each flow location. The values for flow location 1 appear first, while those for flow location 2 appear to the right of those for location 1, etc. Record type 7 is used once for each solute (record type 7 is used NSOLUTE times).

Record types 4–7 repeat for each change in the flow variables. For example, if the flow variables change every 15 minutes (QSTEP = 0.25 hour), these record types will repeat four times for every hour of simulation time. As a result of this repetition, the unsteady flow file may contain a large number of records.

Table 22.   Unsteady flow file — record types 4–7.

[D, Double precision; L, chosen length unit (Section 3.3.1)]

Record type	Input variable	Format	Units	Description
4[1,2]	QLATIN	D	L^3 second^{-1} L^{-1}	Lateral inflow rate from location j-1 to j
5[1,2]	Q	D	L^3 per second	Flow rate at flow location j
6[1,2]	AREA	D	L^2	Main channel area at flow location j
7[1,2,3]	CLATIN	D	moles per liter	Lateral inflow conc. from location j-1 to j

[1]The block of record types 4–7 repeats for each change in the flow variables.

[2]QLATIN, Q, AREA, and CLATIN repeat horizontally for each flow location specified (the values for flow location 1 appear first, the values for location 2 appear to the right of those for flow location 1, etc.).

[3]Record type 7 is used once for each solute (NSOLUTE times).

3.3.6   The MINTEQ Input File

The MINTEQ input file defines the chemical components, reactions, and environmental parameters used within the equilibrium submodel. Unlike the other input files, the MINTEQ input file is not created manually using a text editor; the MINTEQ input file is created using PROTEQ, an adaptation of the problem definition (PRODEF) software distributed with MINTEQ (Allison and others, 1991). Because the MINTEQ input file is generated by the PROTEQ software, only the most common record types are presented here. Additional information on the use of PROTEQ is provided in Section 3.4.3, the model applications (Sections 4.1.4, 4.2.2, 4.3.3, and 4.5.1), and by Allison and others (1991).

Record Types 1–4 — Temperature and Ionic Strength Corrections

Equilibrium constants used within the equilibrium submodel are obtained from a thermodynamic database (Section 3.3.7). Database values are referenced to 25° C and zero ionic strength and should therefore be corrected for ambient conditions. Record types 1–4 (table 23) are used to specify temperature, ionic strength, and the applicable corrections.

The first record type is used to specify the water temperature, TEMPC. This temperature is used to determine temperature-corrected equilibrium constants based on a power function or the van't Hoff equation (Allison and others, 1991). Within OTEQ, temperature is spatially and temporally constant, and TEMPC should be set to a value that is representative of average conditions.

Record type 2 is used to specify the ionic strength option, ISOPT. If ISOPT is set to 0, ionic strength is computed by the equilibrium submodel; if ISOPT is set to 1, fixed ionic strength is assumed. Under the fixed ionic strength option, the ionic strength (TIONS) is specified using record type 3. Fixed ionic strength is often used within OTEQ applications because it reduces the computational demands of the equilibrium submodel. In addition, relatively nonreactive ions such as sodium and chloride that contribute to ionic strength need not be included as transported solutes. Fixing ionic strength has the disadvantage that spatial and temporal variation in ionic strength is not considered; as with TEMPC, a representative value should be used.

The activity coefficient option, KKDAV, is specified using record type 4. If KKDAV is set to 1, the Davies equation is used to develop activity coefficient corrections for the equilibrium constants; if KKDAV is set to 2, the Debye-Huckel equation is used. As noted by Allison and others (1991), Debye-Huckel parameters are not available for all reactions. In cases where the necessary parameters are unavailable, the Davies equation is used.

Record Type 5 — Maximum Iterations

Record type 5 is used to specify ITMAX, the maximum number of Newton-Raphson iterations within the equilibrium submodel.

Record Type 6 — Sorption Option

The sorption option, IADS, is specified using record type 6. Although the equilibrium submodel includes seven sorption algorithms (Allison and others, 1991), only the "diffuse layer model" is implemented within OTEQ (the diffuse layer model is synonymous with the "generalized two layer model" described in Section 2.3.3). If sorption is not being modeled (ISORB=0, parameter file record type 16), IADS is set to 0. When sorption is modeled (ISORB>0), IADS is set to 7 such that the diffuse layer model is invoked.

Record Types 7 and 8 — Surface Definition for Sorption (optional)

Note: Record types 7 and 8 are omitted if sorption is not modeled (IADS=0, record type 6).

As described in Section 2.3.3, sorption may involve static and(or) dynamic surfaces. The number of adsorbing surfaces, NADS, is specified using record type 7. When only a single surface is considered (ISORB=1 or 2, parameter file record type 16), NADS is set to 1. When both static and dynamic surfaces are considered (ISORB=3), NADS equals 2. Record type 8 specifies the specific surface area of each surface, SSA, and is used NADS times.

Record Type 9 — Component Definitions

Record type 9 is used to specify the component identification number (IDX), the total concentration (T), and the log activity guess (GX) for each component. Record type 9 is used once for each solute defined in the parameter file (see record types 12 and 13) and additional times for the sorption components (Section 2.3.3).

Additional Record Types

In addition to the record types presented above, several other record types may be present in the MINTEQ input file. These record types include those needed to specify possible solids, gases at fixed partial pressure, and chemical reactions not within the MINTEQ database files. As with the other record types, these additional records types are created by the problem definition software, PROTEQ (see Section 3.4.3). Examples of these additional record types are given in Section 4.

Table 23. The MINTEQ input file — record types 1–6.

[D, Double precision; I, Integer]

Record type	Input variable	Format	Column	Units	Description
1	TEMPC	D	1–5	degrees Celsius	Temperature
2	ISOPT	I	1–5	—	Ionic strength option (0 or 1)
3	FIONS	D	1–9	moles per liter	Ionic strength
4	KKDAV	I	1–5	—	Activity coefficient option (1 or 2)
5	ITMAX	I	1–5	—	Maximum number of iterations
6	IADS	I	1–5	—	Sorption option (0 or 7)

Table 24. The MINTEQ input file — record types 7 and 8.

[D, Double precision; I, Integer]

Record type	Input variable	Format	Column	Units	Description
7	NADS	I	1–5	—	Number of sorption surfaces (1 or 2)
8[1]	SSA	D	1–13	$meter^2$ per gram	Specific surface area

[1]Record type 8 is used once for each surface (NADS times).

Table 25. The MINTEQ input file — record type 9.

[D, Double precision; I, Integer; Record type 9 is used once for each component]

Input variable	Format	Column	Units[1]	Description
IDX	I	1–7	—	MINTEQ component number
T	D	9–18	moles per liter	Total component concentration
GX	D	20–26	log(moles per liter)	Log free activity guess

[1]Units for T and GX are in terms of moles of sites per gram of sorbent for the sorption components.

3.3.7   The MINTEQ Database Files

The database files distributed with OTEQ are based on the database files distributed with version 3 of MINTEQ (Allison and others, 1991). The version 3 files have been updated to provide consistency with WATEQ (Ball and Nordstrom, 1991) and PHRE-EQC (Parkhurst and Appelo, 1999), as described in Appendix 1. A brief overview of the database files is given here; additional information is provided by Allison and others (1991).

Required Database Files

Several of the MINTEQ database files are required for OTEQ execution. These files include thermo.dbs, type6.dbs, comp.dbs, error.dbs, and analyt.dbs. Three of these files (thermo.dbs, type6.dbs, comp.dbs) are used to define the reactions and thermodynamic data (enthalpy values and equilibrium constants) used within the equilibrium submodel. Error messages used by the equilibrium submodel are contained in error.dbs. A final database, analyt.dbs, contains coefficients for the power function used to develop temperature-corrected equilibrium constants.

Auxiliary Database Files

Several auxiliary database files are used by the problem definition software (PROTEQ) to assist the user in defining specific reactions. Databases gases.dbs and redox.dbs are used to define gases at fixed partial pressure and redox reactions, respectively.

Two additional database files are used to define sorption reactions for the diffuse layer model. The database file feo-dlm.dbs includes 40 reactions between aqueous components and a solid surface composed of hydrous ferric oxide (Dzombak and Morel, 1990). These reactions may be added to the MINTEQ input file using PROTEQ as described in Section 3.4.3. A second file, feo-dlm2.dbs, is identical to feo-dlm.dbs, except that the reactions are defined in terms of a second surface. Reactions for this second surface are added to the MINTEQ input file when concurrent sorption to static and dynamic surfaces is modeled (Section 2.3.3; ISORB=3, parameter file record type 16). Use of the sorption database files is described in Sections 4.3 and 4.5.

3.4   Input File Preparation and Model Execution

Instructions for routine execution of the OTEQ solute transport model under the Unix and Linux operating systems are provided in this subsection. These instructions assume that the user has already installed the software and created work areas (Section 5.3). Prior to running OTEQ, the control, parameter, flow, and MINTEQ input files should be prepared by the user as described below.

3.4.1   Preparation of the Control File

The control file is created using a text editor (vi or emacs, for example) to specify the record types described in Section 3.3.3 (table 1).

3.4.2   Preparation of the Parameter and Flow Files — Use of MINTEQ

As with the control file, the parameter and flow files are prepared using a text editor. The record types comprising these two files are described in Sections 3.3.4 and 3.3.5 (tables 2–22). When modeling simple systems (Section 4.1, for example), specification of the various record types is generally straightforward. Additional considerations come into play when modeling system pH or when using alkalinity to determine dissolved inorganic carbon. In these cases, specification of parameter file record type 28, flow file record type 3 (steady flow), and flow file record type 7 (unsteady flow) requires information obtained from stand-alone MINTEQ runs. Use of MINTEQ to develop these record types is described in detail in Sections 4.2–4.5; execution of MINTEQ is described below.

Prior to executing MINTEQ in stand-alone mode, an input file is created using PRODEF, the problem definition software distributed with MINTEQ (Allison and others, 1991). The PRODEF session is initiated from the MINTEQ work area (Section 5.3.3) by entering the prodef command at the shell prompt:

> prodef

This command will start PRODEF; the user then creates an input file as described by Allison and others (1991). The created input file is then used to conduct a stand-alone MINTEQ run. This run is initiated from the MINTEQ work area by entering the minteq command at the shell prompt:

> minteq

When MINTEQ execution is complete, the output file is inspected and the appropriate record types are set as illustrated in Sections 4.2–4.5.

3.4.3   Preparation of the MINTEQ input file — Use of PROTEQ

The MINTEQ input file used with OTEQ is typically generated using PROTEQ, a modified version of the original problem definition software (PRODEF) distributed with MINTEQ (Allison and others, 1991). During PROTEQ execution, the user is guided through a series of text-based prompts and menus. At the conclusion of the PROTEQ session, a user-generated MINTEQ input file is created.

The PROTEQ session is initiated from the OTEQ work area (Section 5.3.3) by entering the proteq command at the shell prompt:

proteq

This command will start PROTEQ and bring up a prompt that asks the user to specify a name for the MINTEQ input file. The filename specified should be the same as that provided in the control file (Section 3.3.3) and have a maximum length of 20 characters. After specifying the MINTEQ input filename, the user may optionally request that an old file be used as a template. This template or "seed" file is used if the user specifies a filename at the seed file prompt. If a new MINTEQ input file is to be created from scratch, a simple carriage return is entered at the seed file prompt. After the filename prompts, the user is presented with a menu known as Edit Level I.

Edit Level I

Edit Level I is used to set record types 1–5 in the MINTEQ input file (Section 3.3.6). The menu for Edit Level I presents seven options and the corresponding default values. For most OTEQ applications, only menu items 1 and 3 need to be edited by the user. Menu item 1 is used to specify temperature (TEMPC, MINTEQ input file record type 1); menu item 3 is used to define the ionic strength (ISOPT and FIONS, MINTEQ input file record types 2 and 3). The remaining menu items may be left at their default values for most applications.

Edit Level II, Menu Item 1 — Specify Aqueous Components

After leaving Edit Level I, the user is presented with a main menu through which Edit Level II may be accessed. Menu item 1 of Edit Level II allows the user to select aqueous components and assign total concentrations, such that record type 9 of the MINTEQ input file is developed for each of the modeled solutes.

Aqueous components are selected by responding to a series of prompts. After selecting a specific component, the user is prompted for the component's total concentration. Although this total concentration is not used directly by OTEQ, it is used by PROTEQ to develop an initial log activity guess. The total concentration is typically set to the initial solute concentration observed at the upstream boundary of the modeled system (Section 3.3.4, record type 28). This selection process is continued until all of the solutes defined within the parameter file (Section 3.3.4, record types 12 and 13) are defined as aqueous components.

The remaining PROTEQ menus are not required for the most basic OTEQ applications. Additional menu items are needed when precipitation and(or) adsorption reactions are considered, as described below.

Edit Level II, Menu Item 8 — Specify Possible Solids (optional)

If precipitation is being modeled, a possible solid is specified for each precipitate species defined within the parameter file (Section 3.3.4, record type 15). Possible solids are specified by selecting item 8 from the Edit Level II menu.

Edit Level II, Menu Item 3 — Specify Adsorption Model (optional)

If sorption is being modeled (ISORB>0, parameter file record type 16), the relevant sorption information must be entered using menu item 3 of Edit Level II. The first step is to select the diffuse layer model from the menu of available sorption algorithms. After selecting the diffuse layer model, sorption surfaces and reactions are defined.

Defining Sorption Surfaces. Depending on the application, one or two surfaces are defined. When sorption to a static or dynamic surface is considered (ISORB=1 or 2), only a single surface is defined; when sorption to static and dynamic surfaces is considered (ISORB=3), two surfaces are defined. The following conventions apply:

- For a given sorption surface, the first site is the high-affinity site and the second site is the low-affinity site (Section 2.3.3).

- If concurrent sorption onto static and dynamic surfaces is modeled (ISORB=3), surface number one is the static surface and surface number two is the dynamic surface.

A sorption surface is added using menu item 1 from the sorption option menu. The specific surface area (SSA, MINTEQ input file record type 8) of the surface is entered first. The user is then asked to define the first site type associated with the surface (high-affinity site, site type 1), where the concentration of site type 1 is equal to the site density (N_S) divided by the molecular weight of the sorbent (M).[5] A second site type (low-affinity site, site type 2) is then added using menu item 2 from the sorption option menu. The process of defining a surface with two site types is repeated if a second surface is needed (ISORB=3). Example parameters for a surface composed of hydrous ferric oxide (HFO; Dzombak and Morel, 1990) are shown in table 26.

Table 26. Sorption parameters — hydrous ferric oxide (HFO; Dzombak and Morel, 1990).

Parameter	Range	Best Estimate
Specific surface area, SSA [meter^2 per gram HFO]	200–840	600
Molecular weight of HFO, M [gram HFO per mole HFO]	—	89
Site density, N_S, low-affinity site [moles sites per mole HFO]	0.1–0.3	0.2
Site density, N_S, high-affinity site [moles sites per mole HFO]	0.001–0.01	0.005

Defining Sorption Reactions. The final step in setting up the sorption problem is to define the reactions that take place between the aqueous components and the sorption surface. OTEQ applications to date (Broshears and others, 1996; Runkel and others, 1999; Runkel and Kimball, 2002) have assumed that the surface is composed of hydrous ferric oxide (HFO) as defined by Dzombak and Morel (1990). The instructions which follow are therefore specific to the HFO database provided with the equilibrium submodel. As discussed by Allison and others (1991), other user-specified sorption reactions may be defined.

Sorption reactions for the HFO surface are added using menu item 4 from the sorption option menu. This option allows the HFO database (feo-dlm.dbs) to be added to the MINTEQ input file so that the sorption reactions are available to the equilibrium submodel. If two surfaces are modeled (ISORB=3, static and dynamic sorption), menu item 4 is used a second time to add the database for the second surface (feo-dlm2.dbs).

3.4.4 Execution of OTEQ

Before running OTEQ, the required input files (control, parameter, flow, and MINTEQ input files) should be placed in the OTEQ work area (Section 5.3.3). To run OTEQ, enter the following command at the shell prompt:

oteq &

Use of the ampersand (&) after the command name causes the oteq process to run in the background, so that the user has access to the current window during model execution. Status of the oteq process may be monitored using ps and top. In addition, iteration (Sections 2.2.3 and 2.3.4) and time step information is periodically written to Part V of the echo file during program execution. Users may inspect Part V of echo.out during program execution to determine the status of longer OTEQ runs (tail echo.out or more echo.out, for example).

As discussed in Section 5.5.3, OTEQ error messages are written to the screen and the echo output file. Upon completion of the model run, users should inspect echo.out to verify model inputs and to check for execution errors. Specifically, users should inspect the echo file for the following items:

- **Verification of User Input.** Users should inspect echo.out to ensure that the options and parameter values specified in the input files have been read correctly by OTEQ. User input that is not aligned in the correct columns may be read incorrectly by OTEQ, resulting in erroneous simulation results.

- **Verification of the Flow Distribution.** OTEQ sets the volumetric flow rate (Q) in each stream segment based on user-supplied information from the flow file (QSTART, QLATIN, and QLATOUT, for the case of steady flow; Section 3.3.5). Volumetric flow rates at the user-specified print locations are included in Part IV of echo.out. These values should be used to verify that the spatial flow distribution has been correctly initialized by OTEQ.

- **Verification of θ.** The iterative procedure used to implement oxidation/reduction reactions (Section 2.3.4) does not always yield the target fraction specified by the user (θ^{target}; THETA, parameter file record type 24). Users should inspect Part V of echo.out to see how closely the simulated fraction (θ) matches the target fraction (θ^{target}) at the user-specified print locations.

- **Warnings and Error Messages.** Additional warnings and error messages are included in Part V of the echo output file.

---

[5]This value (N_S/M) is used to set T in record type 9 of the MINTEQ input file. Within OTEQ, T is multiplied by the solid concentration (S_C) to determine the total concentration of the sorption component as given by equation 25 (Section 2.3.3).

48 One-Dimensional Transport with Equilibrium Chemistry (OTEQ): A Reactive Transport Model for Streams and Rivers

3.5 Output Analysis

This section describes the mechanics of interpreting the solute concentrations that are written to the solute, solid (precipitate and sorption), and concentration-distance output files. These files are formatted such that the concentration of any phase (eqs. 1 and 42, Section 2) can be calculated from the available output. The discussion includes detailed descriptions of the output file formats (Sections 3.5.1 and 3.5.2), utility programs to reformat the output files (Section 3.5.3), and alternatives for plotting the simulation results (Section 3.5.4).

3.5.1 The Solute and Solid Output Files

As discussed in Section 3.2, the OTEQ solute transport model creates a solute output file for each modeled solute. Solid output files (precipitate and(or) sorption output files) are also created for solutes that are involved in precipitate and(or) sorption reactions (NPRECIPS>0 and(or) ISORB>0, Section 3.3.4). The following paragraphs describe the format of the solute and solid output files. In the discussion that follows, each line is described in terms of a number of fields; each field is 18 characters long.

The format of the solute output file is illustrated in figure 18. Each solute output file contains a time series of total waterborne (mobile) solute concentrations at the user-specified print locations. Total waterborne concentration is given by:

$$T_w = T - P_b - S_b = C + P_w + S_w \tag{52}$$

in the main channel, and

$$T_{ws} = T_s - P_{sb} - S_{sb} = C_s + P_{sw} + S_{sw} \tag{53}$$

in the storage zone (Eqs. 1 and 10, Section 2). The first field of each line of the solute output file is the simulation time, in hours (fig. 18). The next NPRINT fields give the total waterborne solute concentration in the main channel (T_w) at the NPRINT print locations (PRTLOC, Section 3.3.4). If the print option (PRTOPT, Section 3.3.4) is set to 2, the final NPRINT fields give the total waterborne solute concentrations in the storage zone (T_{ws}) at the NPRINT print locations.[6]

Solute Output File

Total Waterborne Solute Concentration

	Main Channel, T_w			Storage Zone, T_{ws}		
Time [hour]	@PRTLOC1	@PRTLOC2...	@PRTLOC$_{NPRINT}$	@PRTLOC1	@PRTLOC2...	@PRTLOC$_{NPRINT}$
1.3000000000E+01	2.5514970966E-06	1.9435631478E-06	2.7531649546E-06	2.8174903149E-06	2.7608659911E-06	2.7888255180E-06
1.3010000000E+01	2.5520255698E-06	1.9423423860E-06	2.7533016031E-06	2.8166289161E-06	2.7591002958E-06	2.7887257154E-06
1.3020000000E+01	2.5525415204E-06	1.9411901809E-06	2.7534373568E-06	2.8157733912E-06	2.7573361855E-06	2.7886265745E-06
1.3030000000E+01	2.5530448754E-06	1.9400188841E-06	2.7535724556E-06	2.8149236633E-06	2.7555737115E-06	2.7885280914E-06
1.3040000000E+01	2.5535358426E-06	1.9387703773E-06	2.7537070356E-06	2.8140796562E-06	2.7538127644E-06	2.7884302625E-06
1.3050000000E+01	2.5540149403E-06	1.9374143319E-06	2.7538410737E-06	2.8132412954E-06	2.7520531376E-06	2.7883330845E-06
1.3060000000E+01	2.5544830222E-06	1.9359409690E-06	2.7539746372E-06	2.8124085095E-06	2.7502945817E-06	2.7882365542E-06
1.3070000000E+01	2.5549413879E-06	1.9343511318E-06	2.7541076393E-06	2.8115812313E-06	2.7485368383E-06	2.7881406683E-06

Figure 18. Example solute output file.

The solid output files include a time series of total solid concentrations (total precipitated or total sorbed) and immobile solid concentrations (immobile precipitated or immobile sorbed). Waterborne solid concentrations (mobile precipitated or mobile sorbed) may be obtained by difference. The format of the precipitate output file is illustrated in figure 19. The first field of each line is the simulation time, in hours (fig. 19). The next NPRINT fields give the total precipitated concentration in the main channel (P) at the NPRINT print locations; a second set of NPRINT fields give the immobile precipitate concentrations in the main channel (P_b) at the NPRINT print locations. Two additional sets of NPRINT fields are present if the print option is set to 2: one set of fields for the total precipitated concentration in the storage zone (P_s) and a second set of fields for the immobile precipitated concentrations in the storage zone (P_{sb}) (these additional sets of fields are not shown in figure 19). The sorption output file is directly analogous to the precipitate output file shown in figure 19; simulated concentrations at the NPRINT print locations are output for total sorbed concentration in the main channel (S), immobile sorbed concentration in the main channel (S_b), total sorbed concentration in the storage zone (S_s), and immobile sorbed concentration in the storage zone (S_{sb}).

---

[6]When modeling pH, one of the solutes represents the component associated with total excess hydrogen (Section 2.3.1). For this solute, simulated values of pH in the main channel and storage zone are output rather than T_w and T_{ws}.

Precipitate Output File

Time [hour]	Total Precipitate Main Channel, P			Immobile Precipitate Main Channel, P_b		
	@PRTLOC1	@PRTLOC2...	@PRTLOC$_{NPRINT}$	@PRTLOC1	@PRTLOC2...	@PRTLOC$_{NPRINT}$
1.3000000000E+01	4.3909141799E-05	5.6662887202E-05	0.0000000000E+00	2.6878194475E-05	4.7376100392E-05	0.0000000000E+00
1.3010000000E+01	4.4017580688E-05	5.6981759332E-05	0.0000000000E+00	2.6985634729E-05	4.7676927765E-05	0.0000000000E+00
1.3020000000E+01	4.4126020648E-05	5.7298998783E-05	0.0000000000E+00	2.7093081255E-05	4.7978293421E-05	0.0000000000E+00
1.3030000000E+01	4.4234462044E-05	5.7615024805E-05	0.0000000000E+00	2.7200534022E-05	4.8280137410E-05	0.0000000000E+00
1.3040000000E+01	4.4342905272E-05	5.7930179528E-05	0.0000000000E+00	2.7307993001E-05	4.8582413386E-05	0.0000000000E+00
1.3050000000E+01	4.4451350471E-05	5.8244723377E-05	0.0000000000E+00	2.7415458167E-05	4.8885085623E-05	0.0000000000E+00
1.3060000000E+01	4.4559797437E-05	5.8558839074E-05	0.0000000000E+00	2.7522929491E-05	4.9188126144E-05	0.0000000000E+00
1.3070000000E+01	4.4668245812E-05	5.8872648588E-05	0.0000000000E+00	2.7630406945E-05	4.9491512334E-05	0.0000000000E+00

Figure 19. Example precipitate output file, for the case of PRTOPT=1.

3.5.2   Concentration-Distance Output Files

Concentration-distance output files are generated when the distance option is set to 1 (DOPT=1, Section 3.3.4). These files contain the spatial profiles of the concentrations included in the solute and solid output files (Section 3.5.1) at the end of the simulation. The format of each concentration-distance output file is identical to the corresponding solute and solid output file (figs. 18 and 19), with the exception of column one where distance is output rather than time. The concentration-distance output files contain one line for each segment in the modeled system.

3.5.3   The Post-Processor, POSTEQ

The solute, solid, and concentration-distance output files described above (Sections 3.5.1 and 3.5.2) may be reformatted using a simple post-processor known as POSTEQ. Using the solute and solid output files, POSTEQ creates time versus concentration files for each solute at each print location. Separate distance versus concentration files are created for each solute when concentration-distance files are available (DOPT=1, Section 3.3.4). These post-processor output files may be plotted along with observed data using plotting utilities such as Xgraph and Grace (Section 3.5.4).

To see a list of available post-processing options, type the following on the command line:

> posteq -h

Post-processing is initiated by typing:

> posteq

Storage zone solute concentrations are included in the post-processed files by adding the -s flag to the command shown above (provided PRTOPT=2, Section 3.3.4).

3.5.4   Plotting Alternatives

Model users have several alternatives for plotting results from the OTEQ solute transport model. One alternative is to use the post-processor described in Section 3.5.3 and the Xgraph plotting utility (Harrison, 1989). An example of this alternative is described here. For the example, assume the name of the solute output file for sulfate is so4.out and one of the print locations is at a distance of 100. In this case, the post-processor creates an output file named so4.out.100.xgr (additional .xgr files are created for the remaining print locations). Simulation results and observed data from the print location at 100 are plotted as follows:

> xgraph -bb -tk -m so4.out.100.xgr obs100.dat

where obs100.dat contains the observed sulfate data. Other plotting alternatives include Grace (http://plasma-gate.weizmann.ac.il/Grace) and various spreadsheet programs (the output files may be read into Excel, for example).

4   Model Applications

In this section, several applications of OTEQ are presented. Each application is designed to illustrate one or more unique features of the model. The reader is encouraged to review all of the applications provided to obtain a complete overview of the model's capabilities. Additional applications of the model are presented elsewhere (Broshears and others, 1996; Runkel and others, 1996b; Runkel and others, 1999; Runkel and Kimball, 2002; Runkel and others, 2007).

Each application has four basic components: (1) a description of the model features that are illustrated in the application, (2) a brief problem statement, (3) example input files, and (4) simulation results. The example input files supplement the description of the input requirements presented in Section 3.3. These example input files are annotated to illustrate how the record types are used within the various files. Due to space limitations, not all of the input files are shown for each application; a complete set of input files is available as part of the software distribution (Section 5.2). Finally, note that all of the applications that follow use meters as the unit of length (L; Section 3.3.1).

4.1   Application 1: Time-Variable Simulation of a Solute Pulse with Precipitation

The primary purpose of this application is to illustrate the format of the OTEQ input files. A secondary purpose is to discuss some of the numerical issues associated with the use of OTEQ.

Chapman (1982) presents a hypothetical example in which a double pulse of solutes is injected into a small stream. A modified form of Chapman's example is considered here, in which a 200-meter stream reach is considered. The stream reach has steady, uniform flow and a cross-sectional area of 0.4 meter^2. The volumetric flow rate is 0.1 meter^3 per second and the dispersion coefficient is 2.0 meter^2 per second. Advection and dispersion are the dominant transport mechanisms, and transient storage is negligible ($\alpha = 0$).

Prior to the injection period, the stream is in equilibrium with background solute concentrations (0.0002 mole per liter) and no solid phases are present on the bed. At time equal to 0.01 hour, a solution of $CaCl_2$ is injected such that the concentration at the upstream boundary (x=0) is 0.3 mole per liter. At time equal to 0.03 hour, this initial injection terminates. Several minutes later, at 0.09 hour, a second injection begins in which Na_2SO_4 is injected; the concentration at the upstream boundary is 0.3 mole per liter. This second injection ends at 0.11 hour.

The double pulse injection is depicted in figure 20. Due to the time interval between the injections, the solute pulses are separated in the upstream regions of the stream. As the pulses travel downstream, transport processes disperse the mass, and mixing of the two pulses occurs. As a result of this mixing, the solution becomes oversaturated with respect to a solid phase, gypsum ($CaSO_4 \cdot 2H_2O$). A precipitate therefore forms and is subject to settling. As the main solute pulse passes, clean waters pass over the streambed, and gypsum that has settled to the streambed redissolves. Simulation of this reactive transport problem requires specification of four solutes (Ca, Cl, Na, and SO_4) and a solid phase (gypsum, $CaSO_4 \cdot 2H_2O$).

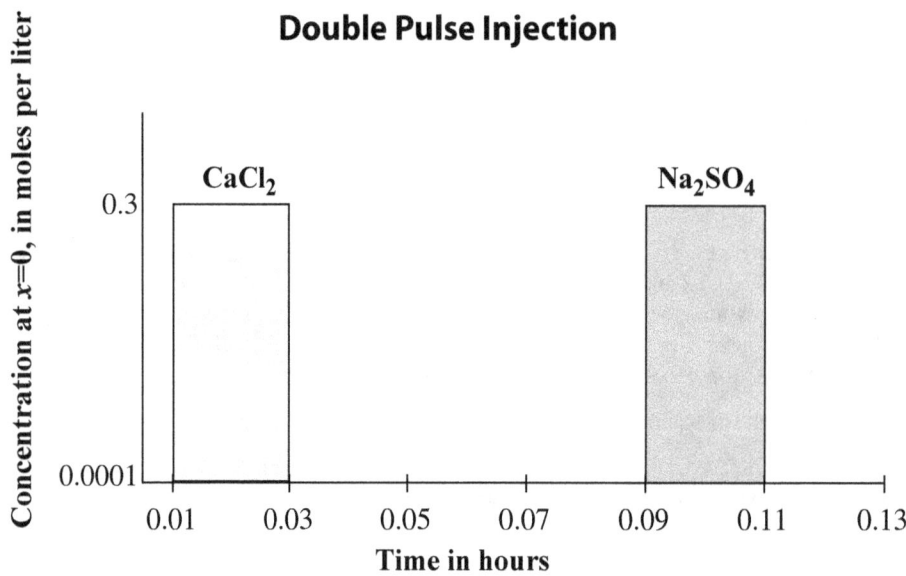

Figure 20.   Upstream boundary condition for double pulse injection. Upstream boundary concentrations for Ca, Cl, Na, and SO_4 are held at background (0.0002 mole per liter) during nonpulse time periods (not shown).

4.1.1 The Control File — Application 1

Example input files corresponding to the double pulse injection are shown below. Figure 21 depicts the control file, where the names of the parameter, MINTEQ, flow, and solute output files are specified. Time-invariant model parameters are read from params.inp, parameters used within the equilibrium submodel are read from minteq.inp, and flow information is read from q.inp (record types 1–3). Simulated solute concentrations are written to four solute output files (record type 4 is used once for each solute). Control files for the remaining applications (Sections 4.2–4.5) are analogous to the control file described here and will not be discussed further.

Control File for Application 1

```
###########################################################################
#
#                OTEQ control file
#
#
# line                 name of the:
# ----                 --------------
# 1                    parameter file
# 2                    MINTEQ input file
# 3                    flow file
# 4 to 3+NSOLUTE    solute output files
#
###########################################################################
1  params.inp
2  minteq.inp
3  q.inp
4  cl.out
4  na.out
4  ca.out
4  so4.out
```

Record Type (label for the numbered lines 1–4 above)

Figure 21. Control file for Application 1.

4.1.2 The Parameter File — Application 1

Figures 22 and 23 depict the parameter file for the double pulse injection. For this simulation, solute concentrations are printed every 0.002 hour (record type 2). The integration time step is 0.001 hours, with the simulation beginning at 0.0 hours and ending at 0.7 hours (record types 3, 4, and 5). The upstream boundary of the network is located at 0.0 meters, and there is a no-flux downstream boundary condition (record types 6 and 7). The relative error tolerance for the sequential iteration procedure is set to 0.001, and storage zone chemistry is not modeled (record types 8 and 9).

The stream is divided into two reaches (record type 10). Record type 11 is used twice to specify reach geometry and reach-specific parameters (dispersion coefficient, storage zone area, exchange coefficient, depth-area coefficient, and exponent).[7] The length of reach one is specified such that the reach ends at the spatial location of interest (200 meters). The second reach extends an additional 100 meters downstream to reduce any error introduced by the downstream boundary condition (see Sections 2.4.2 and 3.3.4). The storage zone exchange coefficient is set to zero (ALPHA=0) due to the negligible effect of transient storage; the storage zone cross-sectional area (AREA2) is set to a non-zero number to avoid division by zero (the value of AREA2 will not affect simulation results, provided ALPHA equals zero). For the settling of solid phases, the channel is assumed to have a rectangular cross section (B1=1.0) that is 1.33 meters wide (A1=1/width=0.75; Section 2.3.6).

---

[7]For this and the subsequent examples, meters are used as the unit of length (L).

Parameter File for Application 1

```
###########################################################################
#                    OTEQ parameter file
###########################################################################
#
# Parameters: TITLE followed by miscellaneous parameters as defined by strings to
#             right of field.
#
# Status     : Required (these fields are always specified)
#
###########################################################################
```

Record Type		
1	Double pulse of calcium chloride and sodium sulfate	
2	0.002	PSTEP [hour]
3	0.001	TSTEP [hour]
4	0.00	TSTART [hour]
5	0.70	TFINAL [hour]
6	0.0	XSTART [L]
7	0.0	DSBOUND [(moles per liter)(L per second)]
8	0.001	TOL
9	1	ICHEM [1=w/o S.Zone chem, 2=w/S.Zone Chem]
10	2	NREACH

```
####################################################################
#
#    Parameters: Reach definition & physical parameters for each reach
#    Status     : Required (these fields are always specified)
#    Note       : NS = NSEG
#
#    Parameters for each reach (J=1,NREACH):
#
#NS RCHLEN  DISP    AREA2   ALPHA   A1     B1
#   |       |       |       |       |      |
####################################################################
```

Record Type	#NS	RCHLEN	DISP	AREA2	ALPHA	A1	B1
11	200	200.	2.0	0.05	0.00	0.75	1.00
11	100	100.	2.0	0.05	0.00	0.75	1.00

```
####################################################################
#
#    Parameters: Solute definition.  NSOLUTE followed by parameters
#                for each solute (J=1,NSOLUTE).
#    Status     : Required (these fields are always specified)
#
####################################################################
```

Record Type					
12	4		NSOLUTE		
13	180	0	0	0	(ID, PFLAG, SFLAG, EFLAG for J=1,NSOLUTE)
13	500	0	0	0	
13	150	1	0	0	
13	732	1	0	0	

```
####################################################################
#
#  Parameters: Precipitate definition.  NPRECIPS followed by
#              parameters for each precipitate (J=1,NPRECIPS).
#  Status     : NPRECIPS required. Additional lines defining the
#              precipitated species required for NPRECIPS > 0
#
#NPRECIPS
#IDPREC  PSETTLE   PSETTLE2
#        |         |
####################################################################
```

Record Type			
14	1		
15	6015001	2.0e-4	0.0

Figure 22. Parameter file for Application 1, record types 1–15.

Parameter File for Application 1, Continued

```
      ############################################################
      #
      #    Parameters: Sorption definition.
      #    Status     : ISORB required. Additional lines required for ISORB > 0
      #
      #    ISORB [0=no sorption,1=static,2=dynamic,3=static & dynamic]
      ############################################################
16    0
      ############################################################
      #
      #    Parameters: Redox definition
      #    Status     : IREDOX required. Additional lines required for IREDOX = 1
      #
      #    IREDOX [0=no redox,1=based on % dissolved]
      #
      ############################################################
22    0
      ############################################################
      #
      #    Parameters: Print definition.
      #    Status     : Required (these fields are always specified)
      #    Note       : NPR = NPRINT
      #
      #NPR IOPT  DOPT PRTOPT
      #    |     |    |
      ############################################################
25    2    0    0    1
26    100.0            (PRTLOC  for I = 1, NPRINT)
26    200.0
      ############################################################
      #
      #    Parameters: Upstream Boundary Condition
      #    Status     : Required (these fields are always specified)
      #    Note       : NBN = NBOUND
      #
      #NBN IBOUND
      #    |
      ############################################################
27    5    1
28    0.00    0.0002    0.0002    0.0001    0.0001
28    0.01    0.6000    0.0002    0.3000    0.0001
28    0.03    0.0002    0.0002    0.0001    0.0001
28    0.09    0.0002    0.6000    0.0001    0.3000
28    0.11    0.0002    0.0002    0.0001    0.0001
      #
      #       Cl        Na        Ca        SO4
```

Figure 23. Parameter file for Application 1, record types 16–28.

Four solutes are modeled (record type 12); record type 13 is used repeatedly to specify the MINTEQ component number and various flags for each solute. Specification of record type 13 for the four solutes follows the order used for naming the solute output files in record type 4 of the control file (Cl, Na, Ca, and SO_4, respectively). The precipitation flag (PFLAG) is set to 1 for Ca and SO_4 to indicate their role in the formation of the gypsum ($CaSO_4 \cdot 2H_2O$) precipitate; the remaining flags in record type 13 are set to zero (sorption reactions and external source/sinks are not considered). The gypsum precipitate is defined using record types 14 (NPRECIPS=1) and 15. Sorption and oxidation/reduction reactions are not considered (ISORB=IREDOX=0; record types 16 and 22); record types 17–21 and 23–24 are therefore omitted.

Simulation results are printed at two locations, 100 and 200 meters downstream (record types 25 and 26). The upstream boundary concentration is in terms of a step concentration profile (fig. 17) representing five time periods (IBOUND=1, NBOUND=5, record type 27). The first, third, and fifth boundary conditions specify the background solute concentrations in the absence of an injection; the second and fourth boundary condition represent the additions of $CaCl_2$ and Na_2SO_4, respectively (record type 28; fig. 20). Upstream boundary concentrations (USBC, record type 28) reflect the stoichiometry of the injectates (there are two moles of Cl in each mole of $CaCl_2$, such that the molar concentration of Cl is twice that of Ca, for example).

4.1.3 The Steady Flow File — Application 1

The flow file for the double pulse injection is shown in figure 24. Steady-flow conditions are specified (QSTEP=0.0, Section 3.3.5), and flow at the upstream boundary is 0.1 meter^3 per second (record types 1 and 2). Record type 3 is used twice to specify the flow variables for each reach. Flow is uniform (QLATIN=0), and the main channel cross-sectional area is 0.4 meter^2 in both reaches. Lateral inflow concentrations (CLATIN) for each solute are arbitrarily set to zero as the magnitude of CLATIN is not of consequence when there is no inflow (QLATIN=0).

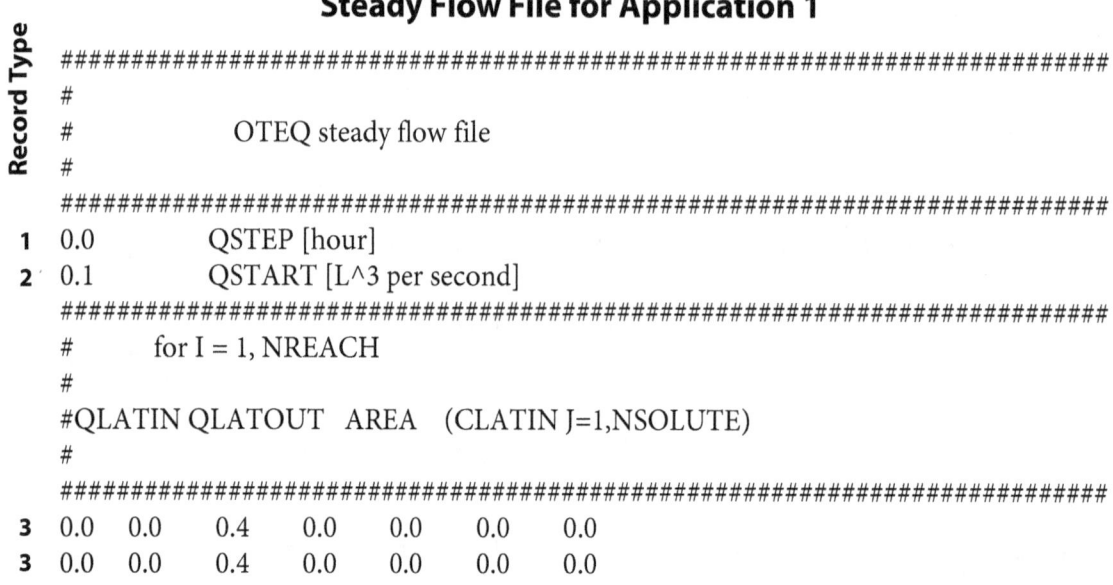

Steady Flow File for Application 1

```
Record Type
  ##############################################################################
  #
  #                    OTEQ steady flow file
  #
  ##############################################################################
1 0.0          QSTEP [hour]
2 0.1          QSTART [L^3 per second]
  ##############################################################################
  #        for I = 1, NREACH
  #
  #QLATIN QLATOUT   AREA   (CLATIN J=1,NSOLUTE)
  #
  ##############################################################################
3 0.0    0.0     0.4     0.0     0.0     0.0     0.0
3 0.0    0.0     0.4     0.0     0.0     0.0     0.0
```

Figure 24. Steady flow file for Application 1.

4.1.4 The MINTEQ Input File — Application 1

The MINTEQ input file created using PROTEQ (Section 3.4.3) is shown in figure 25. Creation of the MINTEQ input file using PROTEQ for the double pulse injection generally follows the description given in Section 3.4.3. Edit Level I is first used to set record types 1–5. Record type 2 sets ISOPT equal to zero so that ionic strength is computed within the equilibrium submodel (examples of fixed, rather than computed, ionic strength are presented in Sections 4.2–4.5). Edit Level II is then used to specify aqueous components (record type 9) corresponding to the four transported solutes (Ca, Cl, Na, and SO_4),[8] with total concentrations equal to the initial concentrations at the upstream boundary (record type 28, parameter file). The MINTEQ component numbers (IDX, record type 9) for the aqueous components must match the component numbers used to define the solutes (ID, record type 13, parameter file). After specifying the aqueous components, menu item 8 of Edit Level II is used to define gypsum ($CaSO_4 \cdot 2H_2O$) as a possible solid. The identification number of the possible solid defined in the MINTEQ input file must match that of the precipitate definition (IDPRECIP, record type 15, parameter file).

[8]In addition to the four transported solutes, a fifth component, H, is automatically added by PROTEQ. Because the precipitation of gypsum ($CaSO_4 \cdot 2H_2O$) is independent of pH, the addition of H is not of consequence; H is therefore not defined as a transported solute in the parameter file. Examples of pH-dependent simulations (where H is a transported component) are provided in Sections 4.2–4.5.

MINTEQ Input File for Application 1

Record Type

```
####################################################################
#
#         MINTEQ input file for use with OTEQ
#
####################################################################
```
1 25.00 | TEMPC
2   0 | ISOPT
3 0.00 | FIONS
4   1 | KKDAV
5 40 | ITMAX
```
#########################################################
#
#         Sorption info - Surface Definition
#
#########################################################
```
6   0 | IADS
```
#########################################################
#
#                    Components
#
#########################################################
```
9   330 0.000E+00 -7.00   /H+1
9   180 0.600E+00 -0.22   /Cl-1
9   500 0.600E+00 -0.22   /Na+1
9   150 0.300E+00 -0.52   /Ca+2
9   732 0.300E+00 -0.52   /SO4-2
```
#########################################################
#
#                    Possible Solids
#
#########################################################
```

     5 1
     6015001   4.8480   -0.2610   /GYPSUM

Figure 25.   MINTEQ input file for Application 1.

4.1.5   Simulation Results — Application 1

Results from the double pulse injection are shown are shown in figure 26, where the simulated concentrations at the second print location (200 meters) are plotted versus time. Because of the time interval between injections, the calcium pulse arrives prior to the sulfate pulse (fig. 26). Shortly after the peak of the calcium pulse, the sulfate pulse arrives. From 0.22 to 0.4 hours, the solution is oversaturated with respect to gypsum due to elevated concentrations of calcium and sulfate. During this time period, gypsum precipitates and mobile precipitate is present in the water column (P_w, fig. 26). This precipitate settles to the streambed, resulting in a gradual increase in immobile precipitate (P_b, fig. 26). As the calcium concentration decreases, the solution becomes undersaturated with respect to gypsum, and the calcium and sulfate that has settled to the bed redissolves (P_b decreases to 0.0 after 0.4 hours).

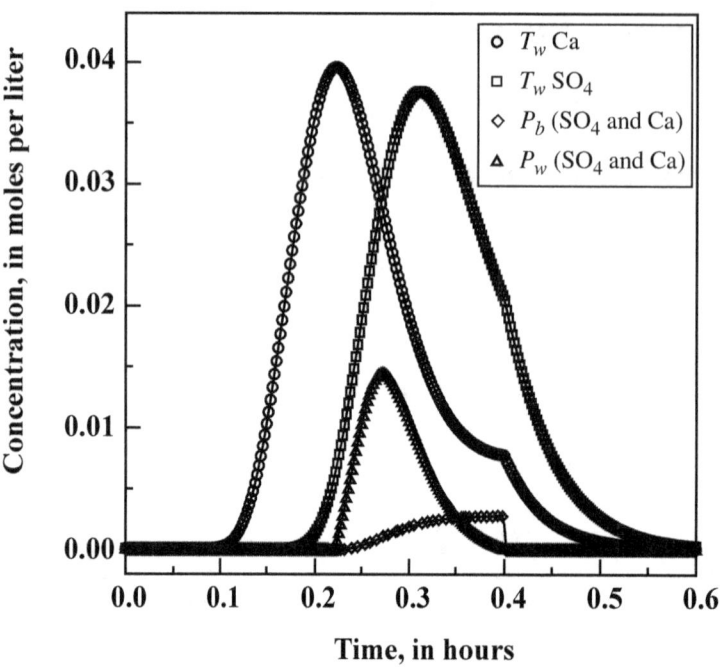

Double Pulse Injection at 200 meters

Figure 26.   Simulated concentrations of calcium (Ca) and sulfate (SO$_4$) at 200 meters, showing the formation of the solid phase.

4.1.6   Numerical Issues — Application 1

The accuracy of the simulation results presented above is highly dependent on the success of the underlying solution techniques. As discussed in Sections 2.2 and 2.4, finite-difference techniques are used to solve the partial differential equations describing transport. Numerical aspects of finite-difference techniques (stability, accuracy, convergence, etc.) are generally well known and are not discussed herein. Model users who are unfamiliar with these basic concepts should consult the detailed descriptions provided by Thomann and Mueller (1987), Chapra and Canale (1988), and Chapra (1997). Two additional numerical issues specific to OTEQ are discussed here. First, a relatively small time step (0.001 hour) is required in this example due to the sharp concentration fronts associated with the double pulse injection (fig. 20). Larger time steps (0.005 hour, for example) result in sequential iteration convergence failures (Section 2.2.3) that affect the accuracy of the simulation. Warning messages are included in Part V of the echo output file when convergence failures occur. In general, the length of the integration time step is application dependent. Steady-state applications such as those presented in Sections 4.4–4.5 often utilize larger time steps than those used for time-variable simulations (Sections 4.1–4.3), for example.

A second numerical issue concerns the treatment of the downstream boundary condition (Section 2.4.2). In this example, reach 2 is a dummy reach 100 meters in length that is included such that the modeled system extends beyond the spatial location of interest (the last print location is at 200 meters; the downstream boundary is an additional 100 meters downstream, at 300 meters). Figure 27 illustrates the effect of the downstream boundary condition and the potential error introduced by placing the downstream boundary too close to last print location. As shown in the figure, a model run with a dummy reach of 100 meters results in a peak total waterborne (dissolved plus mobile precipitate) sulfate concentration of 0.376 mole per liter. When the length of reach 2 is reduced to 50 meters, visually identical results are obtained. Further reductions in the length of reach 2 (10 and 5 meters) result in errors in the peak concentration (fig. 27). As with the integration time step discussed above, the required length of the dummy reach is application dependent. Because of the presence of the dispersion coefficient in equation 49 (Section 2.4.2), applications with large dispersion coefficients will tend to be sensitive to the placement of the downstream boundary condition. Multiple model runs with differing dummy reach lengths should be conducted to assess the error associated with the downstream boundary condition, as shown here.

Effect of Downstream Boundary Condition

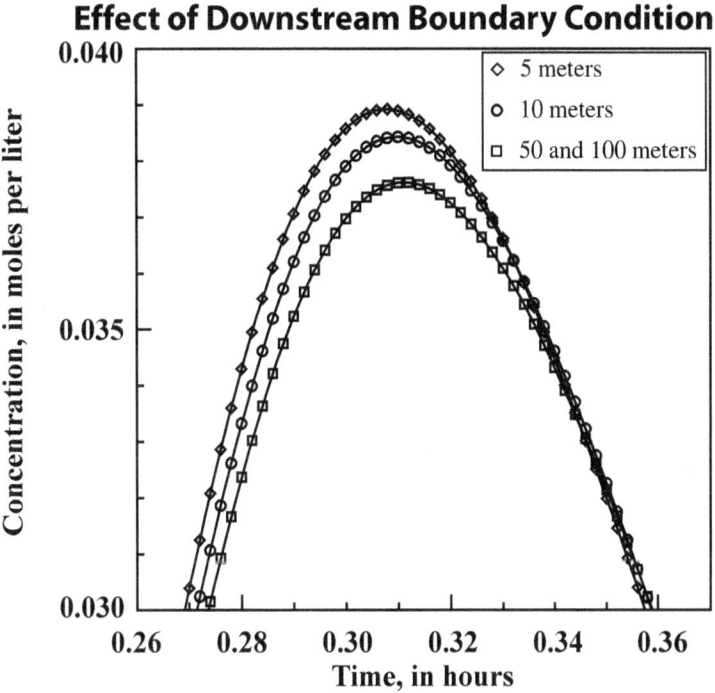

Figure 27. Peak total waterborne (dissolved plus mobile precipitate) sulfate concentration at 200 meters, as a function of reach length for reach 2. The peak concentration is overestimated when reach length is less than 50 meters.

4.2 Application 2: Time-Variable Simulation of pH and pH-Dependent Precipitation

In the foregoing example, a relatively simple geochemical system is presented in which the precipitation of a solid phase is not affected by pH. A more complex problem is considered here, where the precipitation of two solid phases is pH-dependent. The purpose of the example is to illustrate how OTEQ can be used to simulate instream pH, using total excess hydrogen as a transported solute (Section 2.3.1). As shown below, simulation of pH requires specification of total excess hydrogen concentration at the upstream boundary, a quantity that is obtained from a stand-alone MINTEQ run (Section 3.4.2). The stand-alone MINTEQ run is also used to determine the total inorganic carbon concentration at the upstream boundary. A final feature of OTEQ illustrated in this example is the modeling of oxidation/reduction (Section 2.3.4).

Broshears and others (1996) describe a pH-modification experiment conducted in St. Kevin Gulch, a small stream near Leadville, Colorado. During the experiment, a concentrated solution of sodium carbonate (Na_2CO_3) was continuously injected for 5.6 hours. The experiment was designed to raise the acidic waters of St. Kevin Gulch (pH ~3.4) to a pH greater than 5.5. The initial injection rate resulted in a downstream pH of approximately 4.2; after 2.6 hours, the injection rate was increased to further elevate pH (> 5.5). During this second stage of the injection, hydrous oxides of iron and aluminum precipitated in the water column and dissolved concentrations decreased. Decreases in dissolved iron exceeded the available mass of ferric iron, indicating substantial oxidation of ferrous iron at the elevated pH. The pH of St. Kevin Gulch returned to background levels after the injection was terminated, and precipitated mass residing on the streambed redissolved. Additional details on pH modification experiments and St. Kevin Gulch are provided elsewhere (Kimball and others, 1992; Broshears and others, 1996; Runkel and others, 1999; McKnight and others, 2001).

OTEQ simulations of geochemical systems such as St. Kevin Gulch can potentially include numerous solutes corresponding to the complete suite of observed cations and anions. Inclusion of the complete set of solutes results in a large number of chemical species, and solution of the coupled transport/equilibrium problem (Section 2.2.3) becomes computationally intense. In an effort to reduce simulation run time, OTEQ simulations often utilize a reduced set of solutes that adequately describe the problem of interest. Prior to finalizing results, a final OTEQ simulation that includes a more complete set of solutes may be conducted to verify use of the reduced set. In the case considered here, the set of solutes is reduced by noting that some solutes are not involved in chemical reactions. Stand-alone MINTEQ runs, for example, indicate that sodium and chloride exist entirely as free ions in solution (concentrations of the uncomplexed species, Na^+ and Cl^-, account for 100 percent of the total solute concentrations). In addition, reactive solutes such as copper do not appreciably affect the primary constituents of interest (pH, iron, and aluminum). The set of solutes considered in this problem is therefore limited to total excess hydrogen (H), total inorganic carbon (CO_3), ferrous iron (Fe(II)), ferric iron (Fe(III)), aluminum (Al), and sulfate (SO_4).

4.2.1 The Parameter and Flow Files — Application 2

Record types 1–12, 16–21, and 25–26 of the parameter file are very similar to those from Application 1; the reader is referred to Application 1 and the User's Guide (Section 3.3.4) for further information on these record types. The iron and aluminum oxide precipitation caused by the pH modification is simulated by considering the formation of ferrihydrite [Fe(OH)$_3$] and microcrystalline gibbsite [Al(OH)$_3$]. The precipitation flag in record type 13 is therefore set to 1 for Fe(III), Al, and H, to indicate their role in the formation of precipitates.[9] Record types 14 and 15 are used to define the precipitate phases, as in Application 1.

As shown in figure 28, record type 22 of the parameter file is set to 1 to invoke the oxidation/reduction algorithm (Section 2.3.4) so that the oxidation of ferrous iron is considered. Record type 23 is used to specify the two solutes involved in oxidation/reduction, Fe(II) and Fe(III) (MINTEQ component numbers 280 and 281, respectively). Record type 24 is used to specify the fraction of total dissolved iron concentration that is Fe(II) for each reach. The target fraction for each reach (0.80) is based on the fact that ferrous iron comprises 70–90 percent of total dissolved iron during midday hours at St. Kevin Gulch (Kimball and others, 1992; Broshears and others, 1996).

Record Types 22–24 from Application 2 Parameter File

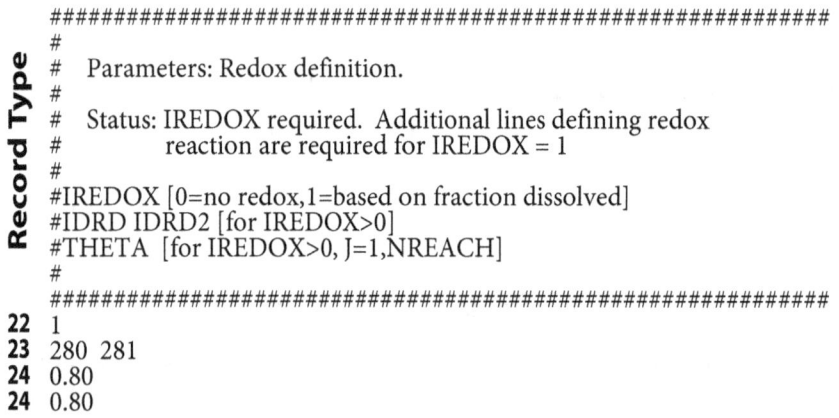

```
############################################################
#
#     Parameters: Redox definition.
#
#     Status: IREDOX required.  Additional lines defining redox
#             reaction are required for IREDOX = 1
#
#IREDOX [0=no redox,1=based on fraction dissolved]
#IDRD IDRD2 [for IREDOX>0]
#THETA  [for IREDOX>0, J=1,NREACH]
#
############################################################
22  1
23  280  281
24  0.80
24  0.80
```

(Record Type labels at left: 22, 23, 24, 24)

Figure 28. Partial listing of the parameter file for Application 2.

Record types 27 and 28 of the parameter file are similar in format to those presented in Application 1, but several differences are of note. First, four boundary conditions are specified, with the first and last boundary conditions corresponding to the pre- and post-injection periods. Boundary conditions two and three correspond to the two different injection rates used to elevate instream pH. Boundary concentrations for Fe(II), Fe(III), Al, and SO$_4$ are set to the observed total waterborne concentrations (T_w) obtained from samples collected above the injection. Because total excess hydrogen (H) has "no direct experimental meaning" (Morel and Morgan, 1972), the boundary concentration for H must be determined from a stand-alone MINTEQ run (Section 3.4.2). Similarly, total inorganic carbon (CO$_3$) is not directly measured and the boundary concentration is determined using a stand-alone run. In the stand-alone MINTEQ run, fixed pH and a fixed partial pressure of atmospheric CO$_2$ (or alternatively, fixed alkalinity) are used in lieu of total component concentrations for H and CO$_3$. These fixed values allow for the calculation of total excess hydrogen (H) and total inorganic carbon (CO$_3$) by mass balance, as described below.

Execution of the stand-alone MINTEQ run proceeds as follows. The PRODEF problem definition software is used to create a MINTEQ input file as described by Allison and others (1991). Edit Level I is used to specify temperature, to fix ionic strength,[10] and to fix pH at 3.6, the observed pH above the injection site. Edit Level II is used to define Fe(II), Fe(III), Al, and SO$_4$ as aqueous components with total concentrations equal to the boundary concentrations specified in record type 28. Edit Level II is also used to specify the fixed partial pressure of CO$_2$ gas, so that total inorganic carbon may be determined from equilibrium with atmo-

---

[9]The precipitation flag for total excess hydrogen (H) is set to 1 because of the presence of hydroxide (OH$^-$) in the precipitates. Precipitation of Al(OH)$_3$ and Fe(OH)$_3$ causes an increase in the excess hydrogen concentration (eq. 15, Section 2.3.1).

[10]The ionic strength specified should correspond to the ionic strength of the water sample from the upstream boundary; that is, the ionic strength that results from the complete suite of cations and anions. Ionic strength is usually determined from a stand-alone MINTEQ run with dissolved concentrations of all the observed solutes, wherein ionic strength is "computed" rather than "fixed."

spheric carbon dioxide.[11] The partial pressure of CO_2 is equal to the percentage of CO_2 in air, corrected for elevation.[12] After creating an input file with PRODEF, MINTEQ is executed and the boundary concentrations (USBC, record type 28) for H and CO_3 are set using the total component concentrations from the equilibrated mass distribution (PART 5 of the MINTEQ output file). For H, the value obtained from the MINTEQ output file is used as the upstream boundary concentration for all four boundary conditions. For CO_3, the MINTEQ-derived value is used to set the pre- and post-injection boundary conditions (boundary conditions one and four). CO_3 boundary concentrations for boundary conditions two and three are set equal to the MINTEQ-derived value plus the CO_3 added by the Na_2CO_3 injection. Input and output files associated with the stand-alone MINTEQ run described here are available as part of the software distribution (Section 5.2).

The steady flow file for this application is similar to that presented for Application 1 (Section 4.1); the reader is referred to Application 1 and the User's Guide (Sections 3.3.5) for further information.

4.2.2   The MINTEQ Input File — Application 2

Creation of the MINTEQ input file using PROTEQ generally follows the description given in Application 1 (Section 4.1.4) and Section 3.4.3. Edit Level I is first used to set record types 1–5. As with the stand-alone MINTEQ run described in Section 4.2.1, ionic strength is fixed at a level corresponding to the complete suite of cations and anions. Edit Level II is used to specify aqueous components (record type 9) corresponding to the six solutes. Ferrihydrite ($Fe(OH)_3$) and microcrystalline gibbsite ($Al(OH)_3$)[13] are defined as possible solids (Edit Level II), with identification numbers that match those used in the parameter file precipitate definition (record type 15, parameter file).

4.2.3   Simulation Results — Application 2

Simulation results for the St. Kevin pH modification are compared to observed data at 24 meters in figure 29. Total dissolved iron (Fe(II)+Fe(III)) and dissolved aluminum concentrations decrease sharply during the second stage of the injection as pH rises above 5.0 (results for pH are similar to those presented by Broshears and others (1996) and are not shown here). For the case of dissolved iron, an initial simulation that does not consider the oxidation of ferrous iron is presented in addition to final simulation results. As shown by the dashed line in figure 29a, the simulation without oxidation overestimates the observed iron concentration. This overestimation occurs because the bulk of the remaining dissolved iron is in the form of Fe(II), a solute that is not part of the $Fe(OH)_3$ solid phase. Final simulation results (solid line, fig. 29a) include the oxidation of Fe(II) to Fe(III), so that additional $Fe(OH)_3$ precipitates, improving the correspondence between simulated and observed iron. For the case of dissolved aluminum (fig. 29b), simulated and observed concentrations are in close agreement during most of the simulation. One exception to this agreement is immediately following the termination of the Na_2CO_3 injection (~15 hours) when the pH returns to background (pH ~3.7). When the pH drops, precipitated mass residing on the streambed redissolves, forming a spike in the simulated and observed aluminum concentrations. Within the equilibrium-based model, this redissolution is immediate, and the spike overestimates observed data. This overestimation and the overestimation of the corresponding iron spike (fig. 29a) suggest a kinetic limitation on the redissolution of precipitated mass that is not considered by the model.

---

[11]In many applications, total inorganic carbon is determined from alkalinity measurements; in the case of St. Kevin Gulch, the acidic water has no alkalinity and CO_3 must be set based on equilibrium with atmospheric CO_2. An example based on alkalinity is shown in Section 4.3.

[12]At sea level, air is 0 03 percent CO_2, such that CO_2 partial pressure equals 0.0003 atm. At 10,000 feet (the approximate elevation of St. Kevin Gulch), atmospheric pressure is approximately 69 percent of the atmospheric pressure at sea level. CO_2 partial pressure at 10,000 feet therefore equals 0.0002 atm. (0.0003 atm. × 0.69).

[13]Another potential solid phase for the precipitation of $Al(OH)_3$ is amorphous $Al(OH)_3$. Microcrystalline gibbsite is used in this example as it provides the best fit to observed data.

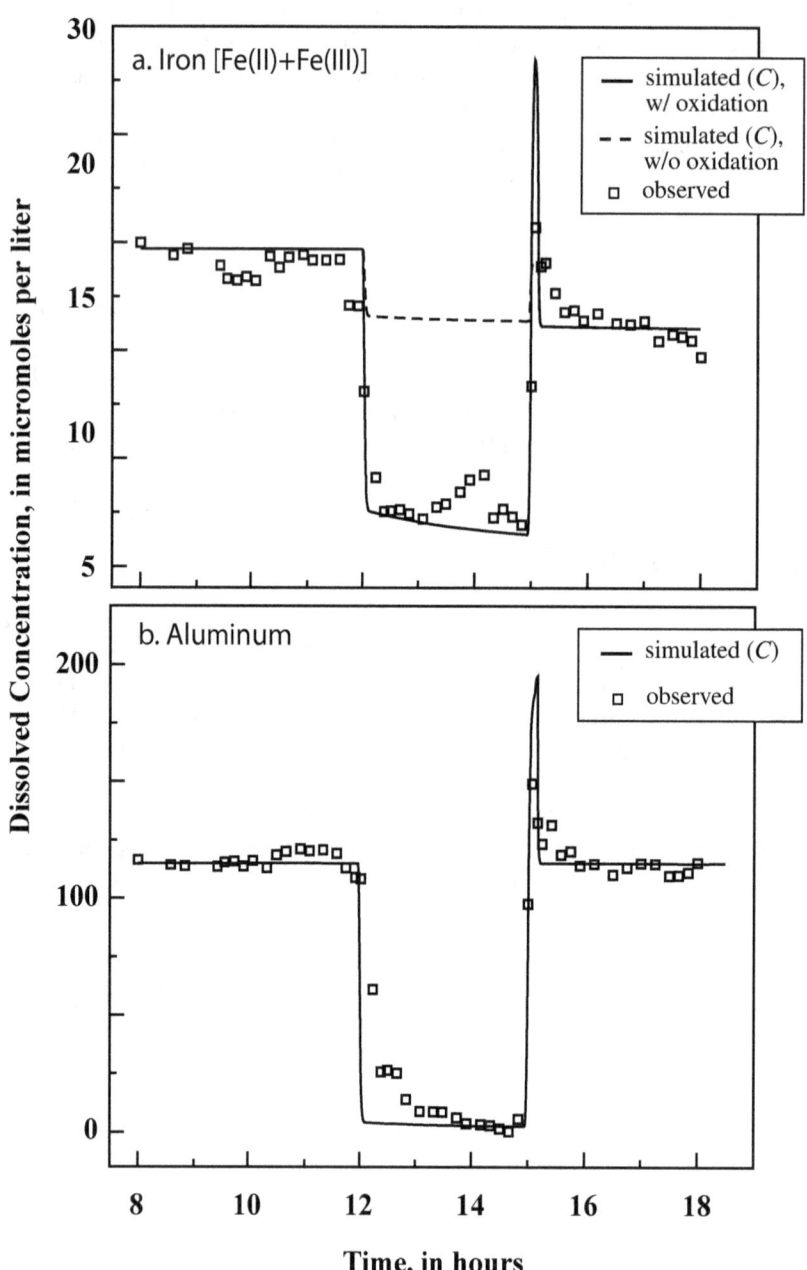

Figure 29.   Simulated and observed concentrations of (a) total [Fe(II)+Fe(III)] dissolved iron and (b) dissolved aluminum at 24 meters.

4.3 Application 3: Time-Variable Simulation of Copper Sorption to the Streambed

The previous two applications illustrate the use of OTEQ to simulate the formation of precipitates in the water column. In this example, OTEQ is used to simulate the sorption of solutes onto the streambed and the subsequent desorption that occurs following passage of a solute pulse. Other concepts introduced in this example include consideration of tributary input, use of the unsteady flow file, and the determination of inorganic carbon from alkalinity. The section concludes with a discussion of numerical errors that may arise when modeling advection-dominated systems.

Copper sulfate ($CuSO_4$) is often used to control algal blooms in lakes and reservoirs used as public water supplies (McKnight, 1981, for example). In this hypothetical example, a 1,000-meter stream is modeled. At approximately 500 meters, a small tributary joins the main stem. The tributary is fed by a small lake that has been treated with copper sulfate. The treatment results in elevated concentrations of Cu and SO_4 at the tributary outlet during a 15-hour period. The 15-hour pulse of Cu and SO_4 is attenuated in the main stem by sorption onto streambed sediments. The main stem is wide and shallow, such that the entire water column is in contact with the streambed and sorption/desorption reactions are in a state of equilibrium. Sorption within the water column is negligible due to a lack of waterborne solid phases. Copper and sulfate desorb from the streambed after passage of the 15-hour pulse. Iron oxides are present in the streambed sediments, such that the sorption database of Dzombak and Morel (1990) is applicable (Sections 2.3.3, 3.3.7, and 3.4.3).

As in Application 2, the number of simulated solutes is reduced by considering only those solutes that are important for the problem at hand. Main stem and tributary waters are circumneutral (pH of 8.2 and 7.0, respectively), such that dissolved metal concentrations (with the exception of copper) are negligible. Simulated solutes include total excess hydrogen (H), total inorganic carbon (CO_3), copper (Cu), sulfate (SO_4), calcium (Ca), and magnesium (Mg). The simulation begins at 0.0 hours and ends at 100.0 hours, with elevated concentrations of Cu and SO_4 entering the main stem via tributary inflow from 5 to 20 hours.

4.3.1 The Parameter File — Application 3

Record types 1–9 are used to specify various simulation parameters and options as described in Application 1 (Section 4.1.2). The tributary input to the main stem is modeled using the segmentation scheme shown in figure 30. The system is simulated using three main reaches plus an additional dummy reach for consideration of the downstream boundary condition (NREACH=4, record type 10). Tributary input is represented by reach 2. Each reach is composed of multiple 10-meter segments (NSEG and RCHLEN, record type 11). Six solutes are defined using record types 12 and 13. The sorption flag in record type 13 is set to 1 for the four solutes with defined reactions in the Dzombak and Morel (1990) database (H, Cu, SO_4, and Ca). No precipitates are modeled (NPRECIPS=0, record type 12).

Sorption to the streambed is modeled using the static surface algorithms (Section 2.3.3). Information relative to the static surface is specified using record types 16–20 (fig. 31). Record type 16 is used to invoke the static surface sorption option (ISORB= 1) and sorption/desorption reactions are in a state of chemical equilibrium (LAMBS=999.0, record type 17). The sorbent solid concentration of the static surface (SOLCON) is specified for each reach (record type 18 is used NREACH times). Record types 19 and 20 are used to specify the initial sorbed concentrations for each solute with defined sorption reactions. Record 19 is first used to specify the MINTEQ component number for SO_4, the first solute involved in sorption (the first solute with SFLAG=1, record type 13). Initial sorbed concentrations (SORBB) for SO_4 are then specified for each reach using record type 20. The block of record types (record type 19 and record type 20 used NREACH times) is then specified for the remaining solutes involved in sorption (H, Cu, and Ca). The initial sorbed concentrations may be determined from stand-alone MINTEQ runs or prior OTEQ simulations.[14]

Specification of the remaining record types in the parameter file is generally similar to that presented for Applications 1 and 2. Two differences are of note. First, due to the larger segment lengths used here (10 meters) and the placement of print locations at segment boundaries (fig. 30), the interpolation option is invoked in record type 25 (IOPT=1). Second, as with Application 2, the upstream boundary concentrations (USBC, record type 28) for total excess hydrogen and total inorganic carbon (CO_3) are based on a stand-alone MINTEQ run (Sections 3.4.2 and 4.2.1). Execution of the stand-alone MINTEQ run is similar to that for Application 2, except that CO_3 is determined from alkalinity rather than equilibrium with atmospheric CO_2. The measured alkalinity concentration is entered when using PRODEF (menu item 6, Edit Level I).

---

[14]For the case considered here, a prior OTEQ simulation was conducted with SORBB set to 0.0 and the tributary inflow concentrations held at background levels. Sorbed concentrations at the completion of the simulation were then used to set SORBB as shown in figure 31. Input and output files associated with the prior OTEQ simulation are available as part of the software distribution (Section 5.2).

Reaches and Segments, Application 3

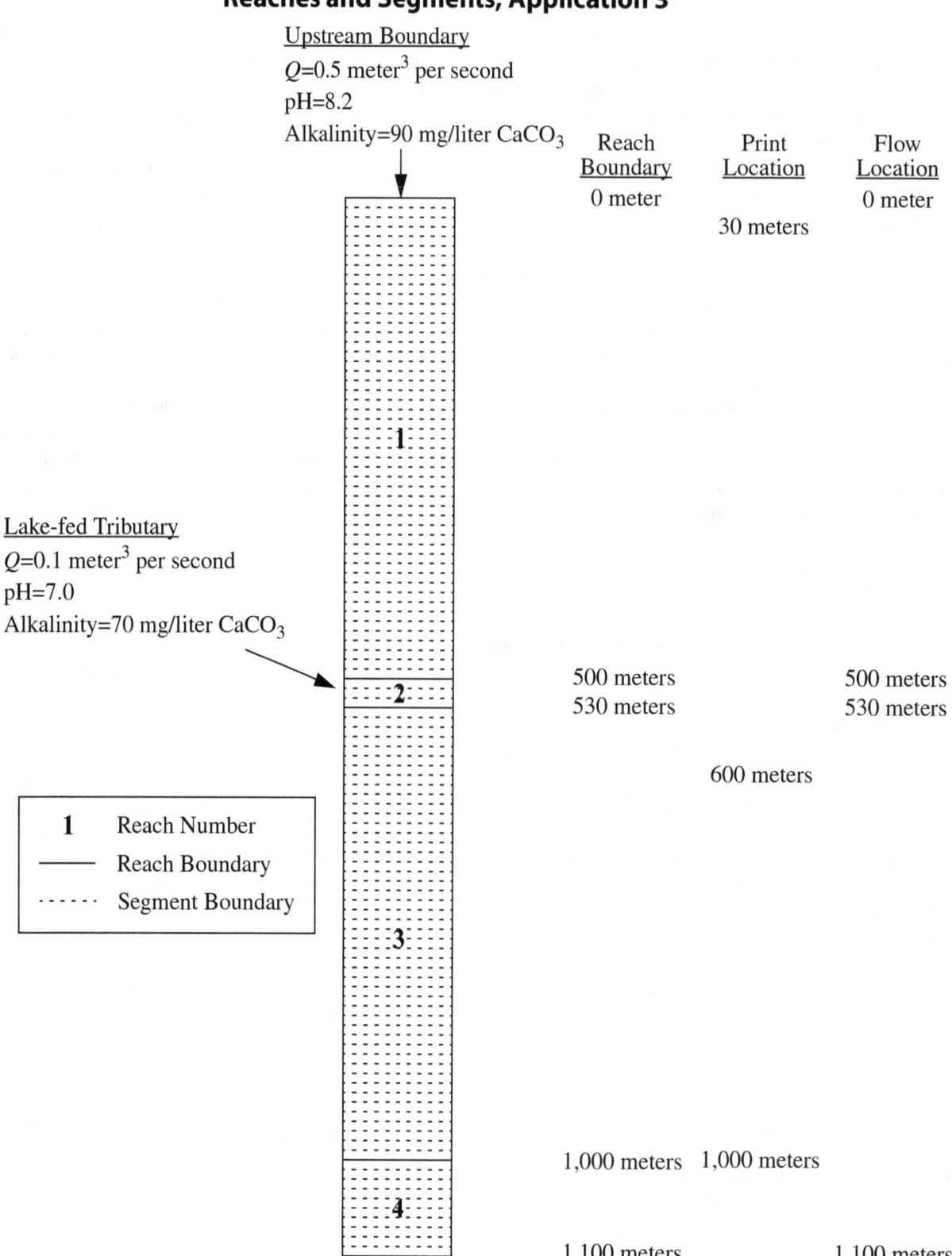

Figure 30.   Segmentation scheme for the hypothetical stream with tributary input. (mg/liter, milligrams per liter)

Record Types 16–20 from Application 3 Parameter File

Record Type

```
################################################################
#
#    Parameters: Sorption definition.
#
#    Status:  ISORB required. Additional lines defining sorption required for ISORB>0
#
################################################################
16  1              ISORB [0=no sorption, 1=static, 2=dynamic, 3=static & dynamic]
17  999.0          LAMBS
18  0.05           SOLCON (J=1,NREACH)
18  0.05
18  0.05
18  0.05
    #
    # SORBB, initial sorbed concentration on bed, SO4
    #
19  732
20  1.0265e-7    0.0
20  2.8759e-7    0.0
20  3.7581e-7    0.0
20  3.7581e-7    0.0
    #
    # SORBB, initial sorbed concentration on bed, H
    #
19  330
20  -2.0803e-5   0.0
20  -1.0875e-5   0.0
20  -9.8279e-6   0.0
20  -9.8279e-6   0.0
    #
    # SORBB, initial sorbed concentration on bed, Cu
    #
19  231
20  4.7210e-7    0.0
20  2.0026e-6    0.0
20  2.5169e-6    0.0
20  2.5169e-6    0.0
    #
    # SORBB, initial sorbed concentration on bed, Ca
    #
19  150
20  1.0188e-5    0.0
20  4.1481e-6    0.0
20  3.2769e-6    0.0
20  3.2769e-6    0.0
```

Figure 31. Partial listing of the parameter file for Application 3.

4.3.2   The Unsteady Flow File — Application 3

The primary purpose of the unsteady flow file is to allow for the simulation of solute transport when the flow regime is unsteady. In this example, a steady-flow regime is considered (flow rates and main channel cross-sectional areas are temporally constant), but the unsteady flow file is needed to accommodate the change in chemistry at the tributary outlet. Use of the unsteady flow file for this purpose is described below; consideration of unsteady-flow regimes is discussed by Runkel (1998) and Runkel and others (1998).

A partial listing of the unsteady flow file is shown in figure 32. Record type 1 specifies QSTEP, the time interval at which the flow variables change. In this example, QSTEP is set to 5 hours, such that the 15-hour period of elevated tributary input is represented by three QSTEPs. Four flow locations are defined using record types 2 and 3. The flow locations are entered in ascending (downstream) order, with the first and last locations at the upstream and downstream ends of the stream network (fig. 30). The second and third flow locations bracket the tributary, so that the changes in flow and cross-sectional area downstream of the tributary outlet are considered.

After defining the flow locations, record types 4–6 are used to set the lateral inflow rate, volumetric flow rate, and main channel cross-sectional area for each flow location. Record type 7 specifies the lateral inflow concentration for each solute at each flow location. The block of record types (record types 4–6 and record type 7 used NSOLUTE times) appears once for every QSTEP hours of simulation time. The first block represents hours 0–5, when the concentrations at the tributary outlet are at background levels. Blocks 2–4 represent hours 5–20 when the tributary concentrations are affected by the copper sulfate treatment. The remaining blocks (20–100 hours) are identical to the first block and represent the return of the tributary concentrations to background levels.

Lateral inflow concentrations (record type 7) at the third flow location are the concentrations associated with the tributary input. Inflow concentrations for Cu, SO_4, Ca, and Mg are set to observed total waterborne concentrations obtained from samples collected at the tributary outlet. Inflow concentrations for H and CO_3 are determined from stand-alone MINTEQ runs in a manner identical to that for the upstream boundary condition (Section 4.3.1).

4.3.3   The MINTEQ Input File — Application 3

Creation of the MINTEQ input file using PROTEQ generally follows the descriptions given in the previous applications and Section 3.4.3. Edit Level I is first used to set record types 1–5, and Edit Level II is used to specify aqueous components corresponding to the six solutes. The relevant sorption information is then entered using menu item 3 of Edit Level II. Within menu item 3, the diffuse layer model is selected from the available sorption algorithms. After selecting the diffuse layer model, the static surface is added using menu item 1 of the sorption option menu. The specific surface area is set to 600 meter^2 per gram, based on the best estimate of Dzombak and Morel (1990) (see Section 3.4.3 and table 26). The high-affinity site type is then defined with a concentration of 5.618×10^{-5} moles per gram. The surface definition is completed by adding a low-affinity site type (menu item 2, sorption option menu) with a concentration of 2.247×10^{-3} moles per gram. Concentrations of the high- and low-affinity site types are based on the best estimates of site density and molecular weight provided by Dzombak and Morel (1990) (see Section 3.4.3 and table 26). The final step in setting up the sorption problem is to define the reactions that take place between the aqueous components and the static surface. Sorption reactions for the static surface are added using menu item 4 from the sorption option menu. This option allows the sorption database (feo-dlm.dbs) to be added to the MINTEQ input file, thereby defining the sorption reactions for the static surface.

Unsteady Flow File for Application 3

Record Type

```
##########################################################################
#    OTEQ unsteady flow file
##########################################################################
1  5.0              QSTEP [hour]
   ########################################################
   #  Flow Locations
   ########################################################
2  4               NFLOW
3     0.0          (FLOWLOC  for I = 1, NFLOW)
3   500.0
3   530.0
3  1100.0
   ########################################################
   #
   # repeating block for each QSTEP:
   #
   #  line 1 - record type 4, QLATIN at each flow location
   #  line 2 - record type 5, Q at each flow location
   #  line 3 - record type 6, AREA at each flow location
   #  lines 4-9 - record type 7, CLATIN for SO4, H, Cu, CO3, Ca, and Mg
   #
   #Flow Loc:
   #  1      2      3       4
   ########################################################
   # 0-5 hours (background conditions)
4  0.00   0.00   3.33e-3   0.00
5  0.50   0.50   0.60      0.60
6  2.00   2.00   2.40      2.40
7  0.00   0.00   6.81e-5   0.00
7  0.00   0.00   2.09e-3   0.00
7  0.00   0.00   1.00e-7   0.00
7  0.00   0.00   1.74e-3   0.00
7  0.00   0.00   5.50e-4   0.00
7  0.00   0.00   7.80e-5   0.00
   # 5-10 hours (elevated Cu and SO4)
4  0.00   0.00   3.33e-3   0.00
5  0.50   0.50   0.60      0.60
6  2.00   2.00   2.40      2.40
7  0.00   0.00   7.00e-5   0.00
7  0.00   0.00   2.08e-3   0.00
7  0.00   0.00   2.00e-6   0.00
7  0.00   0.00   1.74e-3   0.00
7  0.00   0.00   5.50e-4   0.00
7  0.00   0.00   7.80e-5   0.00
   # 10-15 hours (elevated Cu and SO4)
4  0.00   0.00   3.33e-3   0.00
5  0.50   0.50   0.60      0.60
6  2.00   2.00   2.40      2.40
7  0.00   0.00   7.00e-5   0.00
7  0.00   0.00   2.08e-3   0.00
7  0.00   0.00   2.00e-6   0.00
7  0.00   0.00   1.74e-3   0.00
7  0.00   0.00   5.50e-4   0.00
7  0.00   0.00   7.80e-5   0.00
```

(Record types 4–7 repeat for each change in the flow variables.)

Figure 32.   Unsteady flow file for Application 3.

4.3.4   Simulation Results — Application 3

Simulation results for the hypothetical sorption problem are shown in figure 33. Results are shown for a main stem site located 70 meters downstream from the tributary outlet (600 meters) and a second site further downstream (1,000 meters). A conservative simulation (squares, fig. 33) shows the copper concentrations that would result at 600 meters in the absence of sorption. When sorption is modeled, arrival of the copper pulse at 600 meters is delayed by approximately 3 hours (circles, fig. 33). This delay is caused by sorption to streambed sediments located between the tributary outlet (530 meters) and the observation point (600 meters). Copper concentrations at 600 meters begin to increase when all of the upstream streambed sediment is in equilibrium with the elevated tributary input, such that the upstream segments no longer attenuate the copper pulse. The effect of sorption by upstream segments is especially pronounced at 1,000 meters, where the copper arrival is delayed by over 23 hours (diamonds, fig. 33). This extended delay results in a lower peak concentration and a longer period of elevated copper concentrations at 1,000 meters.

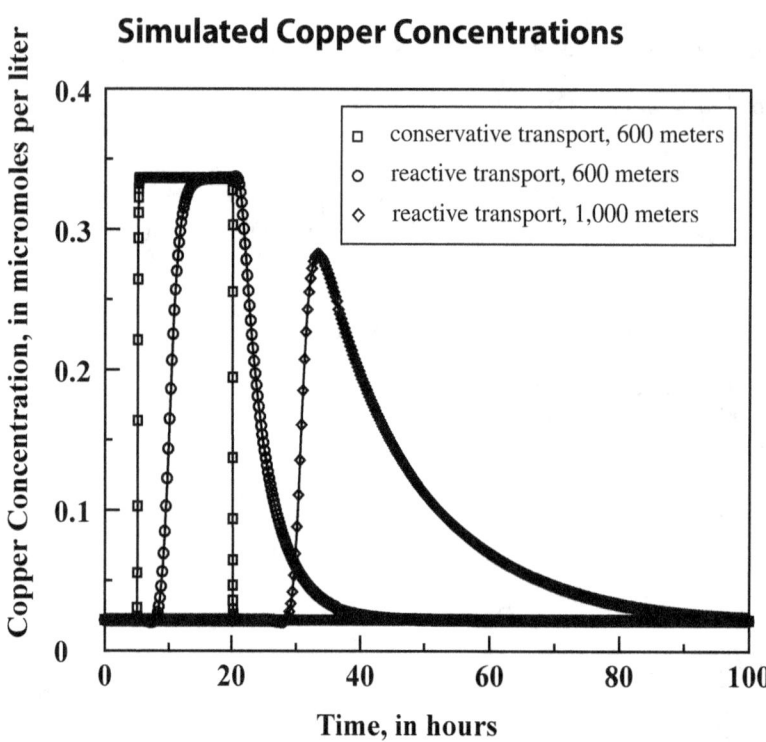

Figure 33.   Copper concentrations at 600 and 1,000 meters, resulting from hypothetical copper sulfate treatment.

4.3.5   Numerical Issues — Application 3

As described in Section 2.4, the governing transport equations within OTEQ are solved using the numerical techniques employed by the OTIS solute transport model (Runkel, 1998). These numerical techniques utilize a Eulerian reference frame in which the finite-difference grid is fixed in space (Runkel and Chapra, 1993, 1994; Runkel and others, 1996a; Runkel, 1998). Eulerian finite-difference methods are known to produce numerical oscillations in the vicinity of sharp concentration fronts when advection-dominated systems are considered (Gray and Pinder, 1976). Fortunately most natural streams and rivers at moderate to low flow are not advection-dominated, due to the mixing induced by dispersion and transient storage. Nevertheless, an illustration of the numerical oscillations that arise when modeling advection-dominated systems is provided below.

Figure 34a shows the spatial profile of copper concentration prior to the copper sulfate treatment (at time=4 hours). The figure corresponds to the simulation described above (Section 4.3.4), where the dispersion coefficient (D) equals 2.0 meters^2 per second and 10-meter segment lengths are used. Copper concentrations increase sharply at 500 meters due to the higher background copper concentrations of the tributary relative to reach 1. Figure 34b illustrates the oscillations that are produced as the system becomes increasingly dominated by advection (as D is decreased). A dispersion coefficient of 1.0 meter^2 per second results in small dip in the copper concentration at 495 meters that precedes the concentration increase due to the tributary. Further reduction of the dispersion coefficient to 0.5 meter^2 per second causes oscillating low and high concentrations (at 475, 485, and 495 meters). Although OTEQ may be inappropriate for some highly advective systems, numerical oscillations can often be avoided by refining the spatial grid. For the case considered here, the oscillations produced when the dispersion coefficient equals 0.5 meter^2 per second can be eliminated by reducing the segment length from 10 to 2.5 meters.

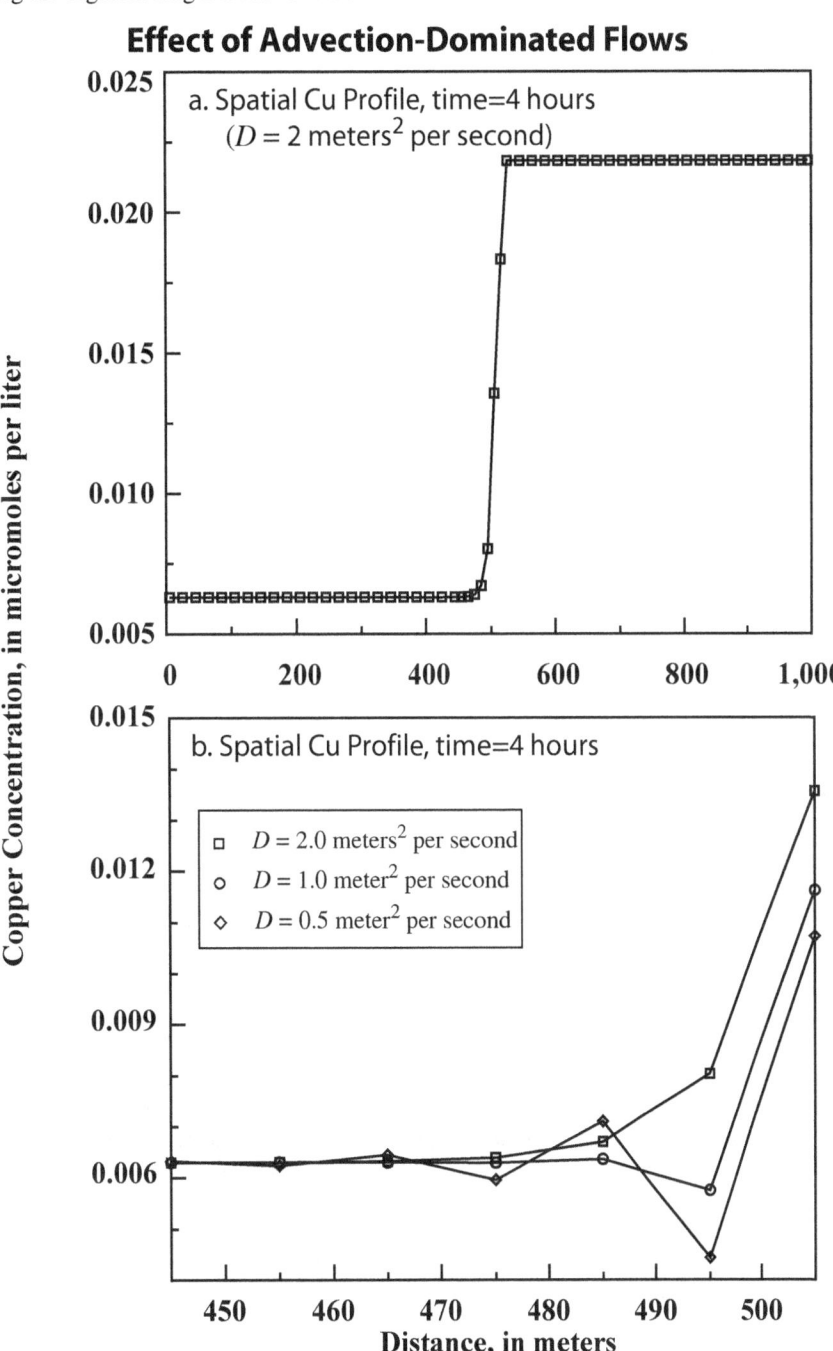

Figure 34.   Spatial profiles of simulated copper concentration at 4 hours: (a) main simulation with D = 2 meter^2 per second (b) additional simulations showing oscillations that arise when D is decreased.

4.4   Application 4: Steady-State Simulation of Existing Conditions and Remedial Action

In addition to the time-variable simulations shown in the previous examples, OTEQ may be used to determine the steady-state solute concentrations that result from a constant loading scenario. Unlike the OTIS solute transport model (Runkel, 1998), a steady-state solution of the governing transport equations (Section 2.2) is not available. OTEQ simulations of steady-state conditions must therefore be based on time-variable runs of sufficient duration to obtain a quasi-steady state, as illustrated below (model "spin-up," Thornton and Rosenbloom, 2005; Section 2.4.3). An additional concept introduced by this application is the use of a calibrated model to simulate potential remedial actions.

Kimball and others (1991) describe a synoptic study conducted in August 1986 at Saint Kevin Gulch, a headwater stream in the Rocky Mountains of Colorado. Saint Kevin Gulch receives acidic, metal-rich waters from a series of springs that emanate from the toe of a large mine dump. Instream metal concentrations increase and pH levels decrease in the vicinity of the dump. Further downstream, a circumneutral tributary (pH 6.5) known as Shingle Mill Gulch provides additional inflow. Metal concentrations decrease downstream of the confluence with Shingle Mill Gulch as the result of dilution and chemical reaction. During the synoptic study, water samples were collected from the acidic springs (363–500 meters), Shingle Mill Gulch (500 meters), several small inflows, and numerous instream sites. Flow rates for the instream sites were computed using the tracer-dilution method; physical transport parameters (D, A, A_S, and α) were estimated using tracer data and a transient storage model (Broshears and others, 1993).

Data from the study provide detailed spatial profiles of flow and concentration that represent steady-state conditions. In this application, OTEQ is used in steady-state mode to evaluate the effect of a potential remedial action. Use of OTEQ in this manner is a two-step process. In the first step, the hydrologic and geochemical processes that influence metal concentrations under existing conditions are quantified. A model of existing conditions is developed using the estimated flows and transport parameters (Broshears and others, 1993) and the inflow chemistry provided by the synoptic sampling. OTEQ is then used to reproduce the observed spatial profiles of pH, iron, and aluminum. In the second step, the calibrated model of existing conditions is modified to reflect the proposed remedial action. In the example considered here, the pH of the acidic springs is increased from 2.7 to 3.6 through the addition of $CaCO_3$. This remedial action is evaluated by conducting an additional simulation in which the inflow concentrations associated with the springs are modified accordingly.

Simulated solutes in this application include total excess hydrogen (H), total inorganic carbon (CO_3), ferrous iron (Fe(II)), ferric iron (Fe(III)), aluminum (Al), and sulfate (SO_4). Precipitation reactions that occur downstream of Shingle Mill Gulch are modeled using ferrihydrite ($Fe(OH)_3$) and microcrystalline gibbsite ($Al(OH)_3$) as possible solids (Sections 4.1.4 and 4.2.2). The set of solutes is intentionally limited to those listed above for illustrative purposes. Simulations of potential remedial actions should include a more complete set of solutes so that unforeseen solute interactions are considered (Runkel and Kimball, 2002).

4.4.1   Quasi-Steady-State Simulations

To model steady-state conditions, a single boundary condition is specified (NBOUND=1, record type 27, parameter file) with upstream boundary concentrations (USBC, record type 28, parameter file) set equal to the observed solute concentrations at 0 meters. In addition, a steady flow file is developed using the estimated flow rates and the observed inflow concentrations (record types 2 and 3, steady flow file). Given a constant loading scenario (a single upstream boundary condition, time-invariant lateral inflow concentrations, and a steady-flow regime), simulated solute concentrations reach steady-state levels after a number of model time steps. Once steady state is achieved, solute concentrations in each model segment provide the spatial concentration profiles corresponding to the steady-state solution. These spatial concentration profiles are obtained by setting the distance option equal to 1 (DOPT, record type 25, parameter file) so that concentration-distance files (Sections 3.2 and 3.5.2) are generated at the end of the model run.

The time to reach steady state is application dependent. Because solute chemistry is modeled as an equilibrium process, only a small number of time steps may be needed to reach steady state. The number of required time steps increases dramatically, however, when settling of waterborne solids (precipitated or sorbed mass), sorption to the streambed, or oxidation/reduction is considered. Model users must therefore inspect the solute output files (Section 3.5.1) to make sure all solute concentrations have reached steady state. If concentrations have not reached steady state, solute concentrations contained in the concentration-distance files will not correspond to steady-state conditions. For the case considered here, solids form below the Shingle Mill confluence and are subject to settling (PSETTLE>0, record type 15, parameter file). Steady-state conditions are verified by inspecting simulation results at the most downstream location of interest (1,804 meters). As shown in figure 35, dissolved ferric iron concentrations at 1,804 meters quickly reach steady-state levels. In contrast, 50 hours of simulation time is required to achieve steady state for total waterborne Fe(III). OTEQ simulations of steady-state conditions at Saint Kevin Gulch are therefore based on 80 hours of simulation time (TSTART=0, TFINAL=80, record types 4 and 5, parameter file).

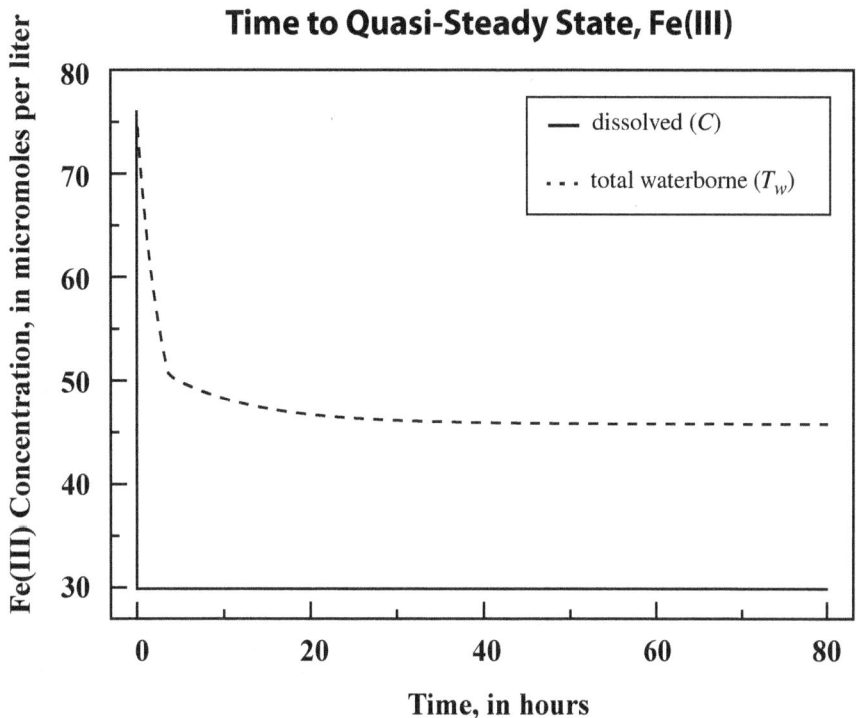

Figure 35. Time required to reach quasi-steady-state conditions. Total waterborne Fe(III) does not reach steady state until 50 hours.

4.4.2   Modeling Existing Conditions and Remediation

The parameter, flow, and MINTEQ input files are similar in format and content to those presented in the preceding applications. The system is modeled using seven reaches. Reach 1 is a short reach (0–26 meters) without lateral inflow (QLATIN=0, record type 3, steady flow file). Flow in reach 2 (26–363 meters) increases slightly due to diffuse inflows that were not sampled during the synoptic. Inflow concentrations (CLATIN, record type 3, steady flow file) in reach 2 are set based on the assumption that the diffuse inflows are chemically similar to the stream. Reach 3 (363–500 meters) includes seven acidic springs that originate at the toe of the mine dump. Direct use of observed data from the springs results in an underestimation of the instream solute concentrations at 500 meters. Lateral inflow concentrations are therefore adjusted upward as part of the calibration procedure. The need for this upward adjustment suggests the presence of a ground-water source with concentrations in excess of the sampled springs. Reach 4 (500–510) is a tributary reach (as in Application 3, Section 4.3.1) corresponding to Shingle Mill Gulch. Lateral inflow concentrations for reach 4 are set using observed data from Shingle Mill Gulch and an alkalinity value of 20 milligrams per liter as $CaCO_3$. Reaches 5 and 6 (510–1,557 meters) introduce additional inflow; inflow chemistry for both reaches is based on observed data for a sampled inflow at 1,281 meters. Reach 7 (1,557–1,904 meters) loses flow to the shallow ground-water system (QLATOUT>0, record type 3, steady flow file). To minimize error associated with the downstream boundary condition (Sections 2.4.2 and 3.3.4), reach 7 extends 100 meters downstream of the last instream sampling location (1,804 meters). As with previous applications, lateral inflow concentrations for H and CO_3 in each reach are set based on a stand-alone MINTEQ run (Sections 3.4.2 and 4.2.1).

Use of the reach chemistry described above results in a simulation that reproduces the observed spatial profiles of pH, iron, and aluminum. This calibrated model of existing conditions can be modified to reflect the changes in solute loading that result from various remedial actions. A hypothetical remediation plan in which the pH of the springs is raised to 3.6 is considered here. To model this remedial action, 5.6×10^{-3} moles per liter of $CaCO_3$ is added to the acidic springs to mimic a small treatment system. A stand-alone MINTEQ run indicates that addition of $CaCO_3$ results in the precipitation of ferrihydrite within the treatment system. Reach 3 inflow chemistry (CLATIN, record type 3, steady flow file) is therefore modified for H, CO_3, and Fe(III) based on dissolved component concentrations from the equilibrated mass distribution (PART 5 of the MINTEQ output file), and OTEQ is used to simulate the effects of remediation. Input and output files associated with the stand-alone MINTEQ runs described here are available as part of the software distribution (Section 5.2).

4.4.3 Simulation Results — Application 4

Simulation results for existing conditions are shown in figure 36. Metal concentrations increase and pH decreases at 363 meters as the acidic springs enter the main stem (fig. 36a–c). Shingle Mill Gulch enters further downstream (500 meters), causing a decrease in metal concentrations and an increase in pH. The processes responsible for the observed decrease in metal concentrations are quantified through model application. For iron, the pH increase caused by Shingle Mill Gulch (fig. 36a) results in the precipitation of ferrihydrite (fig. 36b). The decrease in iron at 500 meters is therefore attributed to dilution (the sharp decrease in total waterborne iron) and chemical precipitation (the difference between total waterborne and dissolved iron). For aluminum, the decrease in concentration is due solely to dilution (fig. 36c; total waterborne and dissolved concentrations are equivalent, indicating an absence of precipitated gibbsite).

The potential effects of remediation on pH, iron, and aluminum are shown in figure 36. When compared to existing conditions, the inflow entering reach 3 (363–500 meters) has a higher pH (3.6 versus 2.7) and lower ferric iron concentration (2.31×10^{-4} versus 3.15×10^{-3} moles per liter) due to treatment of the acidic springs. As a result, instream pH (fig. 36d) increases and iron concentrations decrease (fig. 36e), relative to existing conditions, in the vicinity of the springs. Addition of Shingle Mill water causes further increases in pH and decreases in iron concentrations below 500 meters. As under existing conditions, decreases in iron concentration downstream of Shingle Mill Gulch are attributed to dilution and precipitation. Unlike iron, aluminum concentrations in the reach 3 inflow waters are unaffected by treatment (aluminum remains highly soluble at pH 3.6). Instream aluminum concentrations in the vicinity of the springs are therefore unaffected by remediation (fig. 36f). Nevertheless, remediation does affect aluminum concentrations further downstream, as gibbsite precipitation is an important process downstream of Shingle Mill Gulch where pH exceeds 4.8. In summary, treatment results in the removal of both iron and aluminum, but the location and magnitude of the removal differ for the two solutes. Iron reductions are primarily due to source load reductions (within the treatment system), whereas aluminum reductions are due to instream removal.

4.5 Application 5: Steady-State Simulation of Sorption onto Waterborne Precipitates

In Application 3, OTEQ is used to simulate the time-variable sorption of solutes onto the streambed and the subsequent desorption that occurs following the passage of a solute pulse. As in Application 4, OTEQ is used in this application to determine the steady-state solute concentrations that result from a constant loading scenario. Under steady-state conditions, sorptive surfaces on the streambed are saturated such that sorption to the streambed is negligible (streambed sediments are in equilibrium with the overlying water column). Reactions in the water column, however, may provide a fresh, continuous source of precipitated material that acts as a sorptive surface. The purpose of this application is therefore to illustrate the sorption of solutes onto waterborne precipitates as detailed in Section 2.3.3. In addition, this application illustrates a more detailed approach to specifying total excess hydrogen and total inorganic carbon for the upstream boundary and lateral inflows when waterborne solid phases are present.

Runkel and Kimball (2002) describe a synoptic study conducted in September 1999 at Mineral Creek, a headwater stream near Silverton, Colorado. Mineral Creek receives acidic and metal-rich waters from numerous tributaries, seeps, springs, and ground-water discharge zones located along the water course, resulting in elevated metal concentrations and depressed pH. Data from the synoptic study provide detailed spatial profiles of flow and concentration that represent steady-state conditions. In this application, OTEQ is used in steady-state mode to model a subsection of the 3.5-kilometer study reach described by Runkel and Kimball (2002). The upstream boundary of the subsection is located at 888 meters, where precipitation and sorption reactions result in the formation of waterborne solid phases that are transported downstream. These solid phases are not subject to settling, due to the high velocity of Mineral Creek and the small size of the solid material. The presence of waterborne solid phases at the upstream boundary has important implications for solute transport, as precipitated mass provides a surface for sorption reactions. Observed data at the upstream boundary suggests the presence of precipitated ferric oxides, and the sorption of arsenic and lead onto the precipitated phase. Additional sorption and desorption of metals could occur further downstream as the precipitated phase is transported into reaches with different pH and(or) metal sources (for example, the sorption of copper downstream of 2,400 meters; Runkel and Kimball, 2002). These sorption reactions are modeled using the database of Dzombak and Morel (1990) (Sections 2.3.3, 3.3.7, and 3.4.3) as in Application 3.

Simulated solutes in this application include total excess hydrogen (H), total inorganic carbon (CO_3), ferrous iron (Fe(II)), ferric iron (Fe(III)), aluminum (Al), sulfate (SO_4), and arsenic (As(V)).[15] Precipitation reactions are modeled using ferrihydrite ($Fe(OH)_3$) and microcrystalline gibbsite ($Al(OH)_3$) as possible solids (Sections 4.1.4 and 4.2.2). The set of solutes is intentionally limited to those listed above for illustrative purposes. Simulations that include a more complete set of solutes are presented by Runkel and Kimball (2002). As in Application 4, simulations of steady-state conditions are based on time-variable runs of sufficient duration to obtain a quasi-steady state (Section 4.4.1).

---

[15]Dissolved and sorbed arsenic species in Mineral Creek are assumed to be in the As(V) oxidation state. This assumption is based on the fact that arsenate (As(V)) is the dominant form of arsenic in oxygenated surface waters (Hem, 1985). The reduced form of arsenic (arsenite, As(III)) is not considered here.

St. Kevin Gulch, August 1986 Synoptic

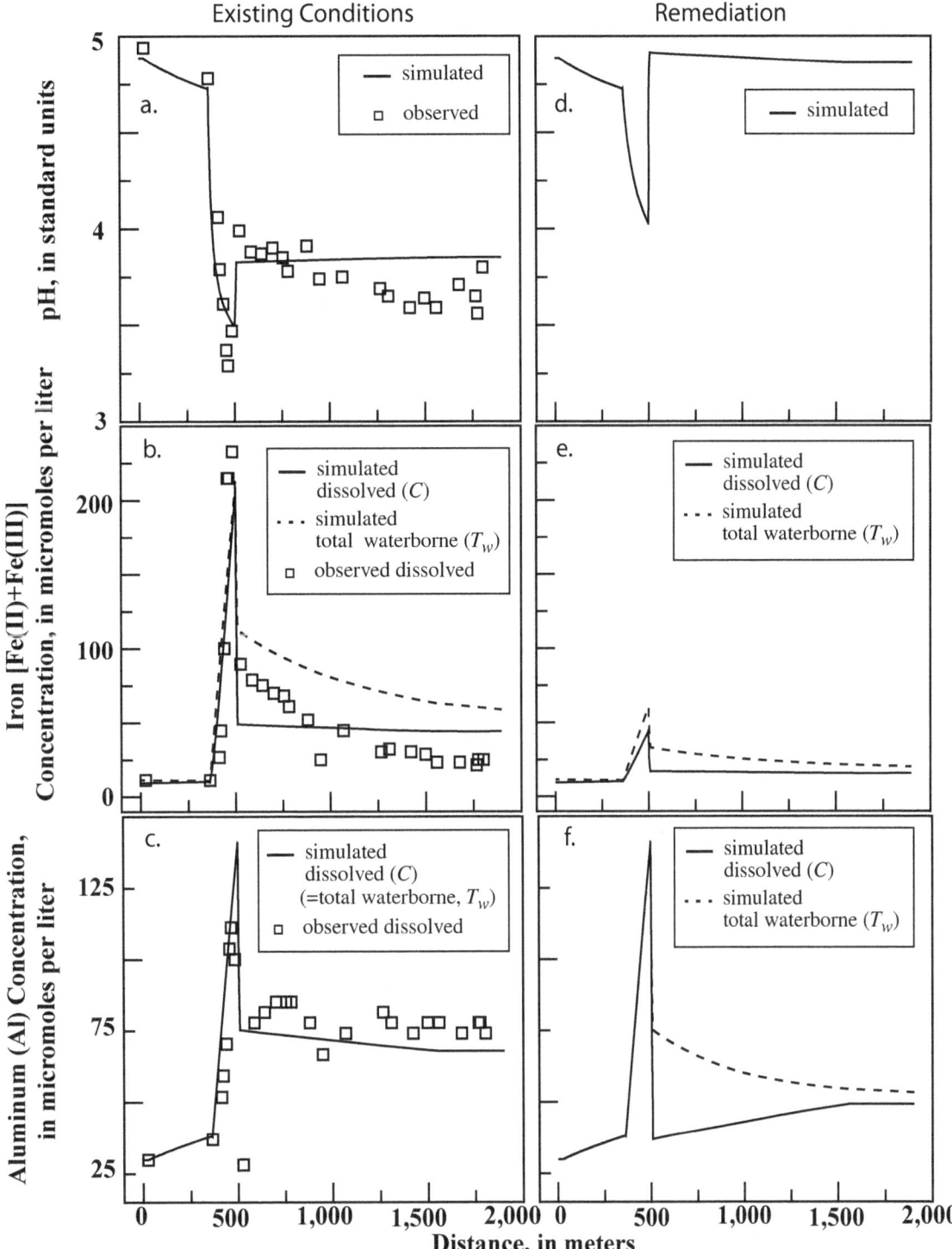

Figure 36. Spatial profiles: (a) pH, existing conditions; (b) iron, existing conditions; (c) aluminum, existing conditions; (d) pH, remediation; (e) iron, remediation; (f) aluminum, remediation.

4.5.1 The Parameter, Flow, and MINTEQ Input Files — Application 5

The parameter, flow, and MINTEQ input files are similar in format and content to those presented in the preceding applications. Within the parameter file, several differences are of note. First, the location of the upstream boundary is set to reflect the subsection of the Runkel and Kimball (2002) study reach that is considered (XSTART=888, record type 6). Second, the lack of solid-phase settling in Mineral Creek is modeled with settling velocities of zero (PSETTLE, record type 15). Due to the lack of settling, solute concentrations reach steady-state values over a relatively short time period; the length of the simulation (TFINAL −TSTART, record types 4 and 5) required to reach quasi-steady state is therefore much less than that for Application 4 (Section 4.4.1). Third, sorption onto waterborne solid phases is modeled using the dynamic surface algorithms (Section 2.3.3). The sorption flag in record type 13 is first set to 1 for the three solutes with defined reactions in the Dzombak and Morel (1990) database (H, SO_4, and As(V)). Record type 16 is then used to invoke the dynamic surface sorption option (ISORB=2; fig. 37); record type 21 is used to specify the MINTEQ component number of the solute that forms the dynamic sorptive surface (IDSORB) and molecular weight of the sorbent (MWSORB). As shown in figure 37, IDSORB is set to 281, the MINTEQ component number for Fe(III), and MWSORB is set to the value provided by Dzombak and Morel (1990). Given this specification, H, SO_4 and As(V) may sorb to precipitated Fe(III) that is present in the water column (ferrihydrite). Fourth, upstream boundary concentrations for H and CO_3 are determined from a series of stand-alone MINTEQ runs as described in Section 4.5.2.

Record Types 16 and 21 from Application 5 Parameter File

```
Record Type

       ###################################################################
       #
       #   Parameters: Sorption definition.
       #
       #   Status:  ISORB required. Additional lines defining sorption required for ISORB>0
       #
       ###################################################################
  16   2           ISORB [0=no sorption, 1=static, 2=dynamic, 3=static & dynamic]
       #
       #   dynamic surface
       #
       # IDSORB   MWSORB
       #            |
       #----------------------------
  21   281         89.0
```

Figure 37. Partial listing of the parameter file for Application 5.

The steady flow file for this application is generally similar to that presented in Application 1 (Section 4.1.3); specification of flow at the upstream boundary and the cross-sectional area of each reach is as described therein. Lateral inflow concentrations for Fe(II), Fe(III), Al, SO_4, and As(V) are set to observed total waterborne concentrations obtained from inflow sampling. Inflow concentrations for H and CO_3 are determined from a series of stand-alone MINTEQ runs as described in Section 4.5.2.

Creation of the MINTEQ input file using PROTEQ generally follows the descriptions given in the previous applications and Section 3.4.3. Edit Level I is used to set record types 1–5; Edit Level II is used to specify the aqueous components and to define ferrihydrite (Fe(OH)$_3$) and microcrystalline gibbsite (Al(OH)$_3$) as possible solids. Specification of the relevant sorption information for the dynamic surface is identical to that presented for Application 3 (Section 4.3.3), with one exception: the high-affinity site type is defined with a concentration of 1.124×10^{-4} moles per gram (0.01/89.0, table 26), reflecting the high sorptive capacity of freshly precipitated iron oxides (Runkel and others, 1999). Following the execution of PROTEQ, the MINTEQ input file is manually edited to change the default value of the surface complexation constant for $H_2AsO_4^-$ to 10.17. This change to the complexation constant is made to improve the correspondence between simulated and observed data for dissolved arsenic (Runkel and Kimball, 2002).

4.5.2 Specification of H and CO₃: The Case of Waterborne Solid Phases

In Applications 2–4, total excess hydrogen (H) and total inorganic carbon (CO_3) concentrations for the upstream boundary and lateral inflows are set based on output from a single stand-alone MINTEQ run. This specification of H and CO_3 concentrations is relatively straightforward due to the absence of waterborne solid phases (total waterborne concentrations are approximately equal to dissolved concentrations at the upstream boundary and in the inflows). The Mineral Creek application, in contrast, has significant waterborne solid concentrations present at the upstream boundary and in some of the inflows. Under these circumstances, a more detailed approach to setting H and CO_3 is required to ensure that the observed pH and solid concentrations are reproduced by the model at the locations of interest (upstream boundary and inflow locations). The more detailed approach is illustrated below for the upstream boundary condition; the same approach is applicable to inflow concentrations when the inflow of interest has waterborne solid phases.

Determination of H and CO3, Step 1

Determination of H and CO_3 concentrations at the upstream boundary is a two-step process. In Step 1, a single stand-alone MINTEQ run is used to determine H and CO_3 concentrations for use in Step 2. The stand-alone run proceeds as follows. PRODEF is used to create a MINTEQ input file. Edit Level I is used to specify temperature, to fix ionic strength, and to fix pH. Edit Level II is used to define Fe(II), Fe(III), Al, SO_4, and As(V)[16] as aqueous components with total concentrations equal to the boundary concentrations specified in record type 28 of the parameter file (the total waterborne concentrations). Total inorganic carbon is set up to be calculated based on equilibrium with atmospheric CO_2 (Edit Level II, Section 4.2.1).[17] After creating an input file, MINTEQ is executed and the concentrations of H and CO_3 listed in PART 5 of the MINTEQ output file (the equilibrated mass distribution) are recorded for use in Step 2.

Determination of H and CO3, Step 2

Step 2 involves a series of stand-alone MINTEQ runs in which pH is estimated, rather than fixed. The goal of the runs is to determine the total excess hydrogen concentration that yields the observed solution pH after the possible solids have precipitated. As before, the MINTEQ input file is created using PRODEF, as follows. Edit Level I is used to specify temperature and to fix ionic strength. Edit Level II is used to define Fe(II), Fe(III), Al, SO_4, and As(V) as aqueous components. Edit Level II is also used to define H and CO_3 as aqueous components, with total concentrations equal to the values determined in Step 1. The input file is completed by using Edit Level II to define the possible solids to be considered within OTEQ (ferrihydrite and microcrystalline gibbsite). After creating an input file, MINTEQ is executed and the estimated pH is compared with the observed pH. Additional MINTEQ runs are then conducted in which the total excess hydrogen concentration is modified to obtain better agreement between observed and estimated pH. Results from this trial and error procedure for the Mineral Creek upstream boundary condition are shown in table 27. As shown in the table, three stand-alone MINTEQ runs are required to reproduce the observed pH at the upstream boundary (4.43). The H concentration used as input for the final stand-alone run and the CO_3 concentration from Step 1 are then used to set the upstream boundary concentrations within the parameter file.

Table 27. Stand-alone MINTEQ runs to reproduce observed pH of 4.43.

MINTEQ run	H concentration input value	Estimated pH
1	-1.019×10^{-4}	4.001
2	-1.519×10^{-4}	4.154
3	-2.019×10^{-4}	4.428

---

[16]The MINTEQ component for As(V) is H_3AsO_4.

[17]For the case of the Mineral Creek upstream boundary, the acidic water has no alkalinity and CO_3 is set based on equilibrium with atmospheric CO_2. Note that when alkalinity is present at the upstream boundary (or in the inflow of interest), PRODEF is used to request the calculation of total inorganic carbon based on the observed alkalinity (Edit Level I, Section 4.3.1).

4.5.3 Simulation Results — Application 5

Simulation results for the Mineral Creek synoptic study are shown in figure 38. The detailed approach for setting the upstream boundary condition described in the previous section results in simulated values that reproduce the observed data at the upstream boundary (888 meters). The simulated precipitation of ferrihydrite results in a waterborne solid phase ($T_w - C$) and instream pH that are comparable to the observed data (fig. 38a and b). The presence of the waterborne solid phase through ferrihydrite precipitation provides a dynamic surface onto which arsenic sorption occurs, reproducing the observed solid phase arsenic ($T_w - C$, fig. 38c). Further downstream, simulated values are generally consistent with observed data, with a close correspondence between observed and simulated values for pH, dissolved iron, and dissolved arsenic. Simulations of total waterborne iron and arsenic underestimate the observed data in reaches 3–7 (1194–1989 meters), suggesting the presence of unsampled inflow waters with high metal concentrations (Runkel and Kimball, 2002).

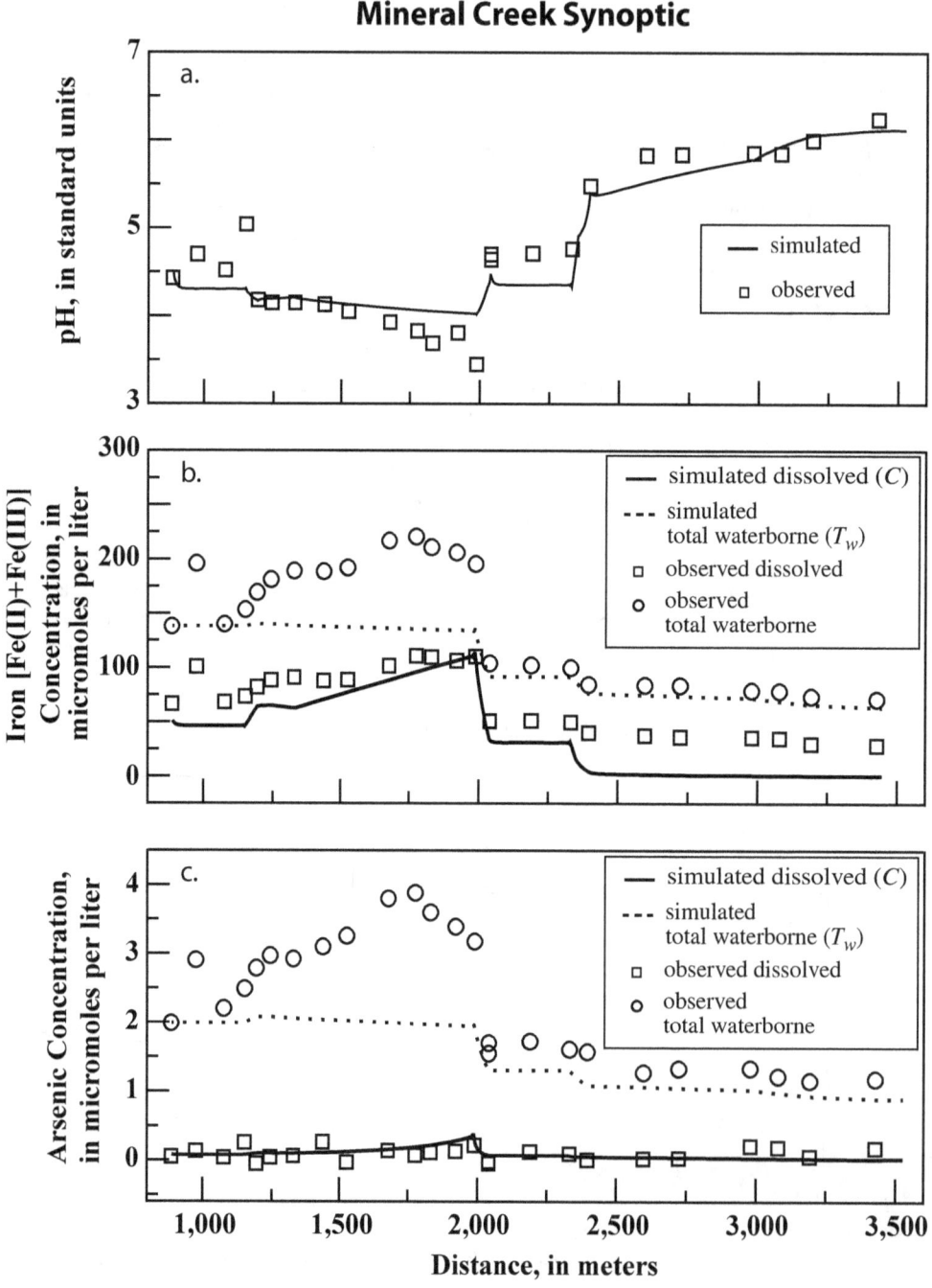

Figure 38. Spatial profiles of simulated and observed: (a) pH, (b) iron, and (c) arsenic.

5   Software Guide

This section provides information on how to obtain and install OTEQ (Sections 5.1–5.4). Additional aspects of the software are described in Section 5.5.

5.1   Supported Platforms

Executable binary files are available for personal computers running Linux and workstations running Solaris. A summary of the supported operating systems and hardware platforms is provided in table 28. Users with other computer systems will need to compile the source code as described in Section 5.4.

Table 28.   Supported systems.

Hardware	Operating System	ARC
Personal computer (Intel x86)	Linux	LIN
Unix workstation (SUN)	Solaris	SOL

Column 3 of table 28 contains ARC, the notation used throughout the remainder of this section to generically denote system architecture. Personal computer users running Linux, for example, should replace the letters "ARC" with "LIN" when following the specific instructions given in Sections 5.2–5.4.

5.2   Software Distribution

OTEQ may be obtained over the Internet at http://water.usgs.gov/software/OTEQ. Model users may download source code, hardware-specific executables, and example input files from the "Download" subpage. Most users will want to download a binary executable and the set of example input files. These files are stored as zipped tar files. A summary of the downloadable files is presented in table 29.

Table 29.   Files to download.

File contents	Zipped tar file
Executable files	oteq.ARC.tar.gz
Source code	oteq.src.tar.gz
Database files, documentation, and example input files	oteq.dde.tar.gz

5.3   Installation

The files described in Section 5.2 may be used to install the OTEQ solute transport model as described here. The installation procedure consists of three steps: (1) creating the OTEQ directory structure, (2) updating the user's path, and (3) creating user work areas. All three steps should be completed on the user's target platform (Linux or Solaris); creation of the OTEQ directory structure within Microsoft Windows is known to cause installation problems.

5.3.1   Creating the OTEQ Directory Structure

The OTEQ directory structure is created using the downloaded files (Section 5.2). To begin, move the downloaded files into the *base-directory*. The *base-directory* is a user-selected directory under which the OTEQ directory tree will be placed (fig. 39). After moving the files, unzip the tar files and extract the directory structure by issuing the following commands:

(1) Unzip the tar files:

      gunzip oteq.dde.tar.gz

      gunzip oteq.src.tar.gz   (if appropriate)

      gunzip oteq.ARC.tar.gz   (if appropriate)[18]

(2) Extract the directory structure:

      tar -xovf oteq.dde.tar

      tar -xovf oteq.src.tar   (if appropriate)

      tar -xovf oteq.ARC.tar   (if appropriate)[19]

5.3.2   Updating the User's Path

After completing the installation procedure, **oteq**, **minteq**, and several other executable files will reside in *base-directory/* **oteq1.40/bin** (fig. 39). A path to these executables is created by editing the user's shell startup files. In most cases, model users will be running one of two types of shells: a C shell or a Bourne shell. Instructions for these two types are provided below. Given the multitude of available shells and the complexity of the startup files, the specific instructions below may not work in all cases (for example, specific shells may use files other than **.cshrc** and **.profile**). Users encountering difficulty with this phase of the installation should contact their system administrator. Also note that changes to the shell startup files will not take effect until the user opens a new window or the startup file is executed using the **source** command.

C Shells

Model users running C shells such as **csh** and **tcsh** can update their path by placing the following lines at the end of their ~/.cshrc file:

               setenv OTEQ_HOME *base-directory*
               set path=($OTEQ_HOME/oteq1.40/bin $path)

where *base-directory* is the user-selected directory where the OTEQ directory tree was placed (Section 5.3.1, fig. 39).

Bourne Shells

Model users running Bourne shells such as **sh**, **ksh**, and **bash** can update their path by placing the following lines at the end of their ~/.profile file:

               OTEQ_HOME=*base-directory*
               export OTEQ_HOME
               PATH=$OTEQ_HOME/oteq1.40/bin:$PATH
               export PATH

where *base-directory* is the user-selected directory where the OTEQ directory tree was placed (Section 5.3.1, fig. 39).

---

[18]Where ARC equals LIN or SOL (see tables 28 and 29).

OTEQ Directory Structure

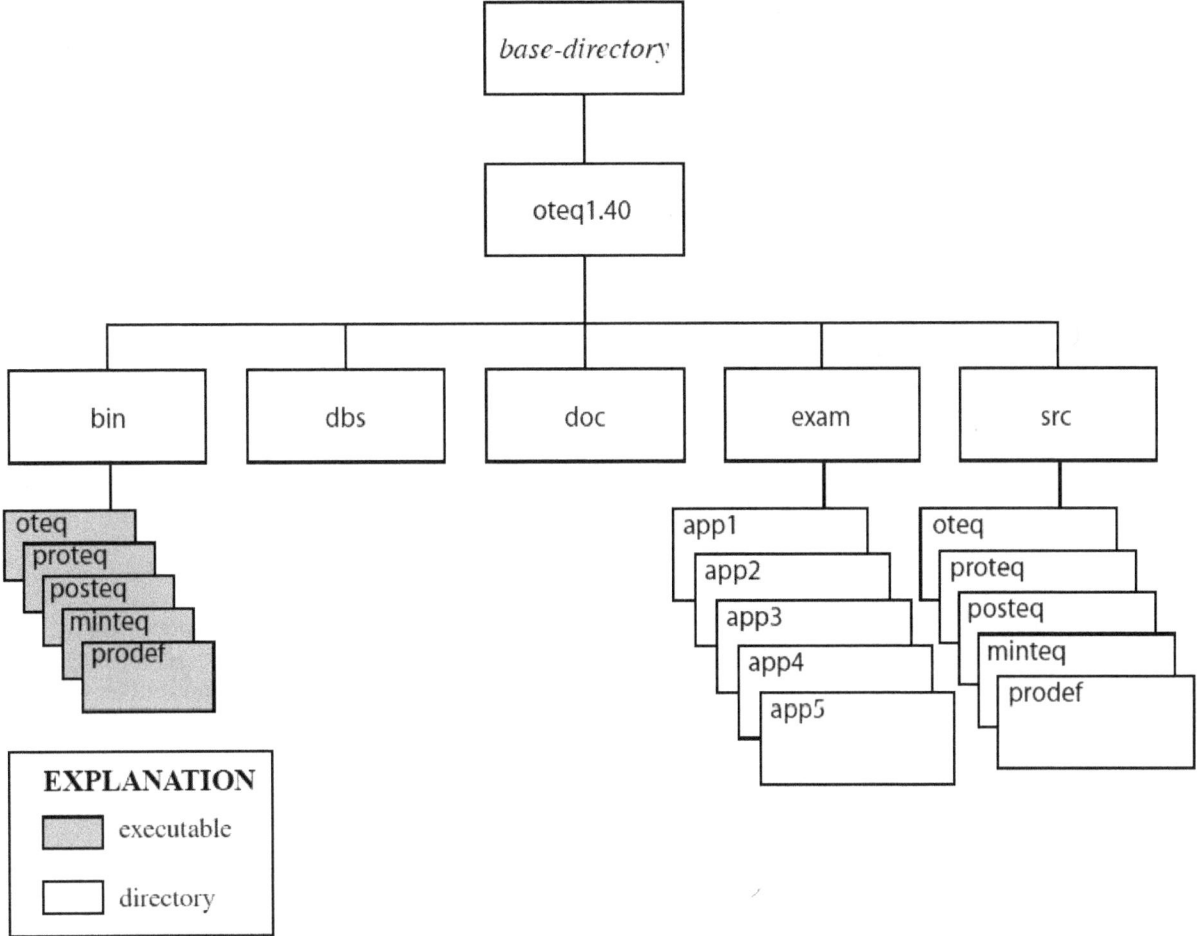

Figure 39. OTEQ directory structure.

5.3.3 Creating User Work Areas

User work areas are user-defined directories from which OTEQ and stand-alone MINTEQ runs are to be executed. Most users will complete multiple model runs, using multiple directories to organize their work. As a result, multiple work areas may be created over time.

Creation of user work areas is illustrated using a simple example. The directory names used below are for example purposes only; model users may name their directories as they see fit. To start, users may want to create a project directory that will contain the OTEQ and MINTEQ work areas. For example, one might create a directory called "redriver" (**mkdir redriver**) for simulations of the Red River. As discussed in Section 3.4.2, one or more stand-alone MINTEQ runs may be required prior to OTEQ execution. A subdirectory and work area for the stand-alone MINTEQ runs may be created for the Red River as follows:

```
cd redriver
mkdir minteq
cd minteq
mkminteqlinks
```

where the final command listed above (**mkminteqlinks**) creates a series of links to the MINTEQ database files. These links point to the actual database files stored in *base-directory*/dbs (fig. 39) and are required for stand-alone MINTEQ execution.[19] Creation of an OTEQ work area proceeds in a similar manner, by issuing the following commands from the **redriver** directory:

```
mkdir run
cd run
mkoteqlinks
```

where the final command listed above (mkoteqlinks) creates a series of links to the MINTEQ database files needed for OTEQ execution.[20] If the user wishes to conduct multiple OTEQ runs, the above process of creating an OTEQ work area may be repeated for other directories (run2, run3, etc., for example).

5.4   Compilation

As discussed in Section 5.1, executable binary files for the OTEQ solute transport model are available for specific hardware platforms and operating systems. As a result, compilation of the source code may not be required. Compilation is required in the following situations:

- Executable binaries are not available for your specific hardware platform/operating system.

- The dimensions of the modeled system exceed the maximum dimensions specified in the include file (Section 5.5.2).

- Modifications have been made to the underlying computer code.

The model is compiled using the make utility. To compile OTEQ, PROTEQ, and POSTEQ, type:

```
make FFLAGS=" " -f makefile.oteq
```

on the command line while in *base-directory*/oteq1.40/src (inserting any applicable compilation flags within the double quotes). To create a specific executable, append ../bin/oteq, ../bin/proteq, or ../bin/posteq to the above command. To compile MINTEQ and PRODEF, type:

```
make -f makefile.minteq
```

on the command line while in *base-directory*/oteq1.40/src. To create a specific executable, append ../bin/minteq or ../bin/prodef to the above command. Note that compilation flags for code optimization are not normally used when compiling MINTEQ and PRODEF, due to issues with the MINVAL subroutine. When compilation is complete, the requested executables will reside in *base-directory*/oteq1.40/bin.

5.5   Software Overview

5.5.1   Model Development

The OTEQ solute transport model is written in ANSI standard Fortran-77. The model has been compiled and tested on two hardware platforms (Section 5.1). Specific development environments used to compile the source code are presented in table 30.

Table 30.   Development environments.

Processor	Operating system	Compiler	Compiler options	ARC
Dual Core Xeon 5050	Mandriva 2007	Portland Group Fortran 77 6.2	-fast -tp px -Bstatic -Mdclchk	LIN
SUN SPARC 10	Solaris 2.6	SPARCompiler Fortran 4.2	-fast -O3 -u	SOL

---

[19] Users may later remove these links using the rmminteqlinks command. Use of rmminteqlinks is for clean up purposes only and is not required (the links do not take up disk space).

[20] Users may later remove these links using the rmoteqlinks command. Use of rmoteqlinks is for clean up purposes only and is not required (the links do not take up disk space).

5.5.2   Include Files

The source code used to develop OTEQ consists of many small subroutines. To facilitate program modification, two include files are used. Use of include files allows program information to be shared between subroutines. This information may be modified by editing the include files rather than each individual routine.

Maximum Dimensions — fmodules.inc

Under Fortran-77, the dimension of each vector and array must be fixed prior to program execution. This requires some knowledge of the maximum dimension for each model parameter. Selection of an appropriate size for each parameter is an important task because excessively small values limit program applicability and excessively large values waste program memory.

To address this problem, the maximum dimensions for the entire model are defined using an include file, fmodules.inc. Increases or decreases in the maximum dimensions are made by editing the include file and compiling the model as described in Section 5.4. Default values for the maximum dimensions are given in table 31. In general, each of the dimensions corresponds to a user-supplied input variable. This correspondence is shown parenthetically in the third column of the table.

When running the model, the input variables may not exceed the maximum values. The number of print locations (see Section 3.3.4 — Record type 25), for example, may not exceed the maximum value given by MAXPRINT. When an input value exceeds its given maximum, program execution is terminated and an error message is issued (Section 5.5.3). At this point the user must increase the appropriate maximum value (by editing fmodules.inc) and recompile the program.

Table 31.   Maximum dimensions and default values from **fmodules.inc**.

Dimension	Default maximum	Maximum number of ...
MAXADS	2	Adsorbing surfaces (NADS)
MAXBOUND	200	Upstream boundary conditions (NBOUND)
MAXCOMP	26	Components
MAXFLOWLOC	30	Flow locations (NFLOW)
MAXPRECIP	10	Precipitates (NPRECIPS)
MAXPRINT	50	Print locations (NPRINT)
MAXREACH	70	Stream reaches (NREACH)
MAXSEG	8000	Stream segments (Σ $NSEG_j$, j=1 to NREACH)
MAXSOLUTE	16	Solutes modeled (NSOLUTE)
MAXSPEC	200	Species

Logical Devices — lda.inc

In Fortran-77, a unit number is assigned to each file used for input and(or) output. These unit numbers, also known as logical device assignments (*ldas*), must be specified for each read and write operation. Program variables used to store the unit numbers are shared between the input and output subroutines using a Fortran common block. This common block is defined in the include file lda.inc.

5.5.3   Error Checking

OTEQ's input subroutines perform several tests to validate the input data. If fatal errors are detected, an error message is written to echo.out and program execution is terminated. The error-checking capabilities of OTEQ are as follows:

- The number of reaches, NREACH, must not exceed the maximum, MAXREACH.
- The number of segments, IMAX, must not exceed the maximum, MAXSEG.
- The number of print locations, NPRINT, must not exceed the maximum, MAXPRINT.
- The number of solutes, NSOLUTE, must not exceed the maximum, MAXSOLUTE.
- The number of upstream boundary conditions, NBOUND, must not exceed the maximum, MAXBOUND.
- The number of flow locations, NFLOW, must not exceed the maximum, MAXFLOWLOC.
- The number of precipitates, NPRECIPS, must not exceed the maximum, MAXPRECIP.
- Sorption information for solutes sorbing to the static surface must be specified in the same order as that used in the initial solute definition.
- The number of precipitates specified in the parameter file, NPRECIPS, must equal the number of possible and finite solids specified in the MINTEQ input file.
- A given print location, $PRTLOC_i$, must lie within the modeled network.
- The flow locations, FLOWLOC, must be entered in ascending (downstream) order.
- The first flow location, $FLOWLOC_1$, must be placed at the upstream boundary. The last flow location must be at or below the downstream boundary.
- The kinetic rate coefficient for sorption, LAMBS, must be equal to 999.0 (equilibrium sorption) or less than the inverse of the integration time step (in seconds).
- The chemistry option, ICHEM, must equal 1 or 2.
- Solutes specified in the parameter file must appear as TYPE I species in the MINTEQ input file.
- Precipitates specified in the parameter file must be defined as a precipitate species in the MINTEQ input file.
- The sorption option must be 0, 1, 2, or 3.
- The component Id specified for the dynamic surface, IDSORB, must be one of the modeled solutes.
- The interpolation option, IOPT, must equal 0 or 1.
- The distance option, DOPT, must equal 0 or 1.
- The print option, PRTOPT, must equal 1 or 2.
- The boundary condition option, IBOUND, must equal 1, 2, or 3.
- The redox option, IREDOX, must equal 0 or 1.
- The component Ids specified for redox, IDRED and IDRED2, must correspond to two of the modeled solutes.
- The sorption option in the MINTEQ input file, IADS, must equal 0 (no sorption) or 7 (diffuse layer sorption).
- The storage zone cross-sectional area, AREA2, must be greater than zero.
- The coefficient of the depth-area power function, A1, must be greater than zero.

References Cited

Alley, W.M., and Smith, P.E., 1982, Distributed routing rainfall-runoff model, Version II, Computer program documentation, user's manual: U.S. Geological Survey Open-File Report 82–344, 205 p.

Allison, J.D., Brown, D.S., and Novo-Gradac, K.J., 1991, MINTEQA2/PRODEFA2, A geochemical assessment model for environmental systems —Version 3.0 user's manual: U.S. Environmental Protection Agency Report EPA/600/3–91/021.

Ball, J.W., and Nordstrom, D.K., 1991, WATEQ4F Users manual with revised thermodynamic data base and test cases for calculating speciation of major, trace and redox elements in natural waters: U.S. Geological Survey Open-File Report 90–129, 185 p.

Bencala, K.E., 1983, Simulation of solute transport in a mountain pool-and-riffle stream with a kinetic mass transfer model for sorption: Water Resources Research, v. 19, no. 3, p. 732–738.

Bencala, K.E., and Walters, R.A., 1983, Simulation of solute transport in a mountain pool-and-riffle stream — A transient storage model: Water Resources Research, v. 19, no. 3, p. 718–724.

Bird, R.B., Stewart, W.E., and Lightfoot, W.E., 1960, Transport phenomena: New York, John Wiley, 780 p.

Broshears, R.E., Bencala, K.E., Kimball, B.A., and McKnight, D.M., 1993, Tracer-dilution experiments and solute-transport simulations for a mountain stream, Saint Kevin Gulch, Colorado: U.S. Geological Survey Water-Resources Investigations Report 92–4081, 18 p.

Broshears, R.E., Runkel, R.L., Kimball, B.A., McKnight, D.M., and Bencala, K.E., 1996, Reactive solute transport in an acidic stream — Experimental pH increase and simulation of controls on pH, aluminum, and iron: Environmental Science and Technology, v. 30, no. 10, p. 3016–3024.

Brown, K.P., and Hosseinipour, E.Z., 1991, New methods for modeling the transport of metals from mineral processing wastes into surface waters: Journal of Environmental Science and Health, v. A 26, no. 2, p. 157–203.

Chapman, B.M., 1982, Numerical simulation of the transport and speciation of nonconservative chemical reactants in rivers: Water Resources Research, v. 18, no. 1, p. 155–167.

Chapra, S.C., 1997, Surface water-quality modeling: New York, McGraw-Hill, 844 p.

Chapra, S.C., and Canale, R.P., 1988, Numerical methods for engineers (2d ed.): New York, McGraw-Hill, 812 p.

Chen, C.W., Leva, D., and Olivieri, A., 1996, Modeling the fate of copper discharged to San Francisco Bay: Journal of Environmental Engineering, v. 122, no. 10, p. 924–934.

Di Toro, D.M., 1976, Combining chemical equilibrium and phytoplankton models - A general methodology, in Modeling Biochemical Processes in Aquatic Ecosystems, Canale, R.P., ed.: Ann Arbor, Michigan, Ann Arbor Science, p. 224–243.

Dzombak, D.A., 1986, Toward a uniform model for the sorption of inorganic ions on hydrous oxides: Massachusetts Institute of Technology, unpublished Ph.D. dissertation, 521 p.

Dzombak, D.A., and Morel, F.M.M., 1990, Surface complexation modeling — Hydrous ferric oxide: New York, John Wiley, 393 p.

Gray, W.G., and Pinder, G.F., 1976, An analysis of the numerical solution of the transport equation: Water Resources Research, v. 12, no. 3, p. 547–555.

Harrison, David, 1989, Xgraph Version 11.3.2: University of California, Berkeley.

Hem, J.D., 1985, Study and interpretation of the chemical characteristics of natural water, 3d ed: U.S. Geological Survey Water-Supply Paper 2254, 264 p.

HydroGeoLogic, Inc., and Allison Geoscience Consultants, Inc., 1999, MINTEQA2/PRODEFA2, A geochemical assessment model for environmental systems — user manual supplement for Version 4.0: Herndon, Virginia, 81 p.

HydroGeoLogic, Inc., 1999, Diffuse-layer sorption reactions for use in MINTEQA2 for HWIR metals and metalloids: Herndon, Virginia, 15 p.

Jobson, H.E., 1989, Users manual for an open-channel streamflow model based on the diffusion analogy: U.S. Geological Survey Water-Resources Investigations Report 89–4133, 73 p.

Keefe, S. H., Barber, L.B., Runkel, R.L., Ryan, J.N., McKnight, D.M., and Wass R.D., 2004, Conservative and reactive solute transport in constructed wetlands: Water Resources Research, v. 40, W01201, doi:10.1029/2003WR002130.

Kimball, B.A., McKnight, D.M., Wetherbee, G.A., and Harnish, R.A., 1992, Mechanisms of iron photoreduction in a metal-rich, acidic stream (St. Kevin Gulch, Colorado, USA): Chemical Geology, v. 96, p. 227–239.

Kimball, B.A., Broshears, R.E., Bencala, K.E., and McKnight, D.M., 1991, Comparison of rates of hydrologic and chemical processes in a stream affected by acid mine drainage, in U.S. Geological Survey Toxic Substances Hydrology Program — Proceedings of the technical meeting, Monterey, California, March 11–15, 1991, Mallard, G.E., and Aronson, D.A., eds.: U.S. Geological Survey Water-Resources Investigations Report 91–4034, p. 407–412.

Kuwabara, J.S., Leland, H.V, and Bencala, K.E., 1984, Copper transport along a Sierra Nevada stream: Journal of Environmental Engineering, v. 110, no. 3, p. 646–655.

McKnight, D.M., 1981, Chemical and biological processes controlling the response of a freshwater ecosystem to copper stress — A field study of the $CuSO_4$ treatment of Mill Pond Reservoir, Burlington, Massachusetts: Limnology and Oceanography, v. 26, no. 3, p. 518–531.

McKnight, D.M., Kimball, B.A., and Bencala, K.E., 1988, Iron photoreduction and oxidation in an acidic mountain stream: Science, v. 240, p. 637–640.

McKnight, D.M., Kimball, B.A., and Runkel, R.L., 2001, pH dependence of iron photoreduction in a Rocky Mountain stream affected by acid mine drainage: Hydrological Processes, v. 15, p. 1979–1992.

Morel, F.M.M., and Hering, J.G., 1993, Principles and applications of aquatic chemistry: New York, Wiley-Interscience, 588 p.

Morel, F., and Morgan, J., 1972, A numerical method for computing equilibria in aqueous chemical systems: Environmental Science and Technology, v. 6, no. 1, p. 58–67.

Nordstrom, D.K., Plummer, L.N., Langmuir, Donald, Busenberg, Eurybiades, May, H.M., Jones, B.F., and Parkhurst, D.L., 1990, Revised chemical equilibrium data for major water-mineral reactions and their limitations, *in* Bassett, R.L., and Melchior, D., eds., Chemical modeling in aqueous systems II: Washington D.C., American Chemical Society Symposium Series 416, chap. 31, p. 398–413.

Nordstrom, D.K., and May, H.M., 1996, Aqueous equilibrium data for mononuclear aluminum species, *in* Sposito, G., ed., The environmental chemistry of aluminum: Boca Raton Fla., CRC Press, Lewis Publishers, chap. 2, p. 39–79.

Nordstrom, D.K., and Archer, D.G., 2002, Arsenic thermodynamic data and environmental geochemistry, *in* Welch, A.H. and Stollenwerk, K.G., eds., Arsenic in ground water: Boston, Kluwer Academic Publishers, chap. 1, p. 1–25.

Parkhurst, D.L., and Appelo, C.A.J., 1999, User's guide to PHREEQC (Version 2) — A computer program for speciation, batchreaction, one-dimensional transport, and inverse geochemical calculations: U.S. Geological Survey Water-Resources Investigations Report 99–4259, 312 p.

Rubin, Jacob, 1983, Transport of reacting solutes in porous media — Relation between mathematical nature of problem formulation and chemical nature of reactions: Water Resources Research, v. 19, no. 5, p. 1231–1252.

Runkel, R.L., 1993, Development and application of an equilibrium-based simulation model for reactive solute transport in small streams: Boulder, University of Colorado, unpublished Ph.D. dissertation, 202 p.

Runkel, R.L., 1998, One-dimensional transport with inflow and storage (OTIS) — A solute transport model for streams and rivers: U.S. Geological Survey Water-Resources Investigations Report 98–4018, 73 p.

Runkel, R.L., 2000, Using OTIS to model solute transport in streams and rivers: U.S. Geological Survey Fact Sheet 138–99, 4 p.

Runkel, R.L., Bencala, K.E., Broshears, R.E., and Chapra, S.C., 1996a, Reactive solute transport in streams 1. Development of an equilibrium-based model: Water Resources Research, v. 32, no. 2, p. 409–418.

Runkel, R.L., and Chapra, S.C., 1993, An efficient numerical solution of the transient storage equations for solute transport in small streams: Water Resources Research, v. 29, no. 1, p. 211–215.

Runkel, R.L., and Chapra, S.C., 1994, Reply to comment on "An efficient numerical solution of the transient storage equations for solute transport in small streams": Water Resources Research, v. 30, no. 10, p. 2863–2865.

Runkel, R.L., and Kimball, B.A., 2002, Evaluating remedial alternatives for an acid mine drainage stream — Application of a reactive transport model: Environmental Science and Technology, v. 36, no. 5, p. 1093–1101.

Runkel, R.L., Kimball, B.A., McKnight, D.M., and Bencala, K.E., 1999, Reactive solute transport in streams — A surface complexation approach for trace metal sorption: Water Resources Research, v. 35, no. 12, p. 3829–3840.

Runkel, R.L., Kimball, B.A., Walton-Day, K., and Verplanck, P.L., 2007, A simulation-based approach for estimating premining water quality—Red Mountain Creek, Colorado: Applied Geochemistry, v. 22, no. 9, p. 1899–1918.

Runkel, R.L., McKnight, D.M., and Andrews, E.D., 1998, Analysis of transient storage subject to unsteady flow — Diel variation in an Antarctic stream: Journal of the North American Benthological Society, v. 17, no. 2, p. 143–154.

Runkel, R.L., McKnight, D.M., Bencala, K.E., and Chapra, S.C., 1996b, Reactive solute transport in streams 2. Simulation of a pH modification experiment: Water Resources Research, v. 32, no. 2, p. 419–430.

Thomann, R.V., and Mueller, J.A., 1987, Principles of surface water quality modeling and control: New York, Harper & Row, 644 p.

Thornton, P.E., and Rosenbloom, N.A., 2005, Ecosystem model spin-up — Estimating steady state conditions in a coupled terrestrial carbon and nitrogen cycle model: Ecological Modelling, v. 189, no. 1–2, p. 25–48.

Westall, J.C., Zachary, J.L., and Morel, F.M.M., 1976, MINEQL — A computer program for the calculation of chemical equilibrium composition in aqueous systems: Massachusetts Institute of Technology, Department of Civil Engineering, Tech. Note 18.

Yeh, G.T., and Tripathi, V.S., 1989, A critical evaluation of recent developments in hydrogeochemical transport models of reactive multichemical components: Water Resources Research, v. 25, no. 1, p. 93–108.

Yeh, G.T., and Tripathi, V.S., 1991, A model for simulating transport of reactive multispecies components — Model development and demonstration: Water Resources Research, v. 27, no. 12, p. 3075–3094.

Glossary

Summary of variables used in Section 2. The fundamental units of Length [L] and Time [T] are used herein.

A main channel cross-sectional area [L^2]

A_S storage zone cross-sectional area [L^2]

C concentration of a component's dissolved phase [moles per liter]

C_s storage zone concentration of a component's dissolved phase [moles per liter]

C^{tot} sum of all dissolved species for the two components involved in oxidation/reduction

$\hat{C}$ main channel concentration of an arbitrary phase [moles per liter]

$\hat{C}_{bc}$ fixed concentration at the upstream boundary [moles per liter]

$\hat{C}_L$ lateral inflow concentration of the arbitrary phase [moles per liter]

$\hat{C}_S$ storage zone concentration of an arbitrary phase [moles per liter]

CC $\exp(-\Psi F/RT_a)$

D dispersion coefficient [L^2T^{-1}]

F Faraday constant [96,485 coulomb mole^{-1}]

H^+ hydrogen ion

K equilibrium constant

K^{int} intrinsic surface complexation constant

$L()$ transport operator

M molecular weight of the sorbent [gram sorbent per mole sorbent]

M number of complexed species

M^{2+} divalent cation

N number of stream segments

N_S site density [moles of sites per mole sorbent]

P concentration of a component's precipitate phase (P_w+P_b) [moles per liter]

P_b concentration of a component's immobile precipitate phase [moles per liter]

P_s storage zone concentration of a component's precipitate phase ($P_{sw}+P_{sb}$) [moles per liter]

P_{sb} storage zone concentration of a component's immobile precipitate phase [moles per liter]

P_{sw} storage zone concentration of a component's mobile precipitate phase [moles per liter]

P_w concentration of a component's mobile precipitate phase [moles per liter]

P^{Me} total precipitate concentration for the specified hydrous metal oxide

P_b^{Me} immobile precipitate concentration of a specified hydrous metal oxide

P_w^{Me} mobile precipitate concentration of a specified hydrous metal oxide

Q volumetric flow rate [L^3T^{-1}]

R molar gas constant [8.314 joules mole^{-1} K^{-1}]

S concentration of a component's sorbed phase (S_w+S_b) [moles per liter]

S_b concentration of a component's immobile sorbed phase [moles per liter]

S_s storage zone concentration of a component's sorbed phase ($S_{sw}+S_{sb}$) [moles per liter]

S_{sb} storage zone concentration of a component's immobile sorbed phase [moles per liter]

S_{sw}	storage zone concentration of a component's mobile sorbed phase [moles per liter]
S_w	concentration of a component's mobile sorbed phase [moles per liter]
S_A	specific surface area [meter^2 per gram sorbent]
S_C	concentration of the sorptive solid [gram sorbent per liter]
S_1	sorbed concentration associated with Pool 1 [moles per liter]
S_2	sorbed concentration associated with Pool 2 [moles per liter]
S_3	sorbed concentration associated with Pool 3 [moles per liter]
SOH	uncharged surface hydroxyl group
SOM^+	positively charged surface species
T	total component concentration [moles per liter]
T_a	absolute temperature [K]
T_s	total storage zone component concentration [moles per liter]
T_H	total component concentration for excess hydrogen [moles per liter]
Z	valence of a symmetrical electrolyte
a	coefficient of the depth-area power function
a_i	stoichiometric coefficient of the component in the *ith* complexed species
a_m	stoichiometric coefficient of the component in the *mth* precipitated species
b	exponent of the depth-area power function
c	concentration of the uncomplexed component species [moles per liter]
c_e	molar electrolyte concentration
d_1	effective settling depth in main channel [L]
d_2	effective settling depth in storage zone [L]
f_b	source/sink term for dissolution from the immobile substrate [moles per liter T^{-1}]
f_{bm}	source/sink term for dissolution of immobile precipitated species m [moles per liter T^{-1}]
f_{sb}	source/sink term for dissolution from the storage zone immobile substrate [moles per liter T^{-1}]
f_w	source/sink term for precipitation/dissolution from the water column [moles per liter T^{-1}]
g_b	source/sink term for sorption/desorption from the immobile substrate [moles per liter T^{-1}]
g_{sb}	source/sink term for sorption/desorption from the storage zone immobile substrate [moles per liter T^{-1}]
g_w	source/sink term for sorption/desorption from the water column [moles per liter T^{-1}]
i	subscript denoting stream segment
k	superscript denoting the iteration number for sequential iteration
m	superscript denoting the iteration number of the oxidation/reduction scheme
n	superscript denoting time level (n denotes an initial time, $n+1$ denotes an advanced time)
np	number of solid precipitate species for the current component
p_{bm}	immobile precipitate concentration for precipitated species m [moles per liter]
p_m	total precipitate concentration for precipitated species m [moles per liter]
q_{LIN}	lateral inflow rate [L^{3}T^{-1}L^{-1}]
s_{ext}	source/sink term representing external gains and losses [moles per liter T^{-1}]
s_{sext}	source/sink term representing external gains and losses for the storage zone

	[moles per liter T^{-1}]
t	time [T]
v_1	main channel settling velocity [LT^{-1}]
v_{1m}	settling velocity for precipitated species m [LT^{-1}]
v_2	storage zone settling velocity [LT^{-1}]
w	channel width [L]
x	distance [L]
x_i	concentration of the *ith* complexed species [moles per liter]
*	superscript denoting the equilibrium concentration in the absence of a kinetic limitation
α	storage zone exchange coefficient [T^{-1}]
Δt	integration time step [T]
Δx	stream segment length [L]
ε	dielectric constant of water
ε_o	permittivity of free space [8.876×10^{-12} coulomb volt^{-1} meter^{-1}]
σ	net surface charge density [coulomb meter^{-2}]
σ_e	relative error tolerance for sequential iteration procedure
Γ	fraction of the equilibrium quantity that is allowed to sorb/desorb during the current time step
Ψ	surface potential [volts]
θ^{target}	fraction of the total dissolved concentration that is associated with the first component involved in oxidation/reduction

Appendix 1. Modifications to the MINTEQ Database

The database files distributed with OTEQ are based on the database files distributed with version 3 of MINTEQ (Allison and others, 1991). The version 3 files have been updated to provide consistency with more recent information. The resultant database files are generally consistent with the databases of WATEQ (Ball and Nordstrom, 1991) and PHREEQC (Parkhurst and Appelo, 1999), and the published data of Nordstrom and others (1990), Nordstrom and May (1996), and Nordstrom and Archer (2002). Details specific to each of the updated database files are discussed in this appendix. Although the updated files are thought to reflect the best available information, OTEQ users should note that:

"logK's and enthalpies of reaction have been taken from various literature sources. No systematic attempt has been made to determine the aqueous model that was used to develop the initial logK's or whether the aqueous models deCEned by the current data base CEes are consistent with the original experimental data. The data base CEes provided in the program should be considered to be preliminary. Careful selection of aqueous species and thermodynamic data is left to the users of the program" (Parkhurst and Appelo, 1999).

Given this potential uncertainty in the logK values and enthalpies of reaction, OTEQ users are encouraged to apply formal sensitivity analysis techniques to their model applications. For a discussion of sensitivity analyses and a simple example showing the effects of reaction enthalpy, see Runkel and others (2007).

Aqueous Species - thermo.dbs

Enthalpy values and equilibrium constants (logKs) contained in MINTEQ's thermo.dbs file were compared with values contained in wateq4f.dat, a database file distributed with PHREEQC (Parkhurst and Appelo, 1999). Values for many aqueous species were identical, as shown in table 32. Enthalpy and(or) logK values in MINTEQ's thermo.dbs file differed from those present in wateq4f.dat for 33 aqueous species; OTEQ's version of thermo.dbs has therefore been changed to use the wateq4f.dat values (table 33). In addition to providing for consistency with PHREEQC's database, many of these changes result in values that are closer to those provided in the revised database distributed with MINTEQ version 4 (table 33). Enthalpy and logK values for aqueous species not listed in tables 32 and 33 have not been checked against wateq4f.dat; values for unlisted species remain at the original values from version 3 of MINTEQ.

Mineral Species - type6.dbs

Enthalpy and logK values contained in MINTEQ's type6.dbs file were compared with values contained in wateq4f.dat, a database file distributed with PHREEQC (Parkhurst and Appelo, 1999). Values for many mineral species were identical, as shown in table 34. Enthalpy and(or) logK values in MINTEQ's type6.dbs file differed from those present in wateq4f.dat for 36 mineral species (table 35). OTEQ's version of type6.dbs has not been changed to reflect these differences, however; default enthalpy and logK values remain at the original values from version 3 of MINTEQ. OTEQ users may easily change enthalpy and logK values (to values from wateq4f.dat or other sources) when adding possible solids using PRODEF or PROTEQ (Sections 3.4.2 and 3.4.3). Enthalpy and logK values for mineral species not listed in tables 34 and 35 have not been checked against wateq4f.dat; values for unlisted species remain at the original values from version 3 of MINTEQ.

Sorption to HFO - feo-dlm.dbs and feo-dlm2.dbs

The auxiliary database file for sorption of inorganic ions onto hydrous ferric oxide (feo-dlm.dbs) distributed with version 3 of MINTEQ (Allison and others, 1991) is based on the doctoral dissertation of Dzombak (1986). The auxiliary database files distributed with OTEQ (feo-dlm.dbs and feo-dlm2.dbs, Section 3.3.7) have been updated to incorporate more recent data from Dzombak and Morel (1990). The updated files include 40 reactions and equilibrium constants that are consistent with version 4 of MINTEQ (HydroGeoLogic Inc., 1999) and PHREEQC (Parkhurst and Appelo, 1999).

Table 32. Aqueous species for which MINTEQ version 3 enthalpy and logK values are identical to those in wateq4f.dat, a database distributed with PHREEQC.

MINTEQ ID	Species	Enthalpy [kilocalorie per mole]	LogK
201300	AgBr	0.000	4.240
201301	AgBr2-	0.000	7.280
201302	AgBr3-2	0.000	8.710
201800	AgCL	-2.680	3.270
201801	AgCL2-	-3.930	5.270
201802	AgCL3-2	0.000	5.290
201803	AgCL4-3	0.000	5.510
202700	AgF	-2.830	0.360
203300	AgOH	0.000	-12.000
203301	Ag(OH)2-	0.000	-24.000
203800	AgI	0.000	6.600
203801	AgI2-	0.000	10.680
203802	AgI3-2	-27.030	13.370
203803	AgI4-3	0.000	14.080
204910	Ag(NO2)2-	0.000	2.220
204920	AgNO3	0.000	-0.290
207300	AgHS	0.000	14.050
207301	Ag(HS)2-	0.000	18.450
207302	Ag(S4)2-3	0.000	0.991
207303	AgS4S5-3	0.000	0.680
207304	Ag(HS)S4-2	0.000	10.431
207320	AgSO4-	1.490	1.290
902700	BF(OH)3-	1.850	-0.399
1505800	CaHPO4	-0.230	15.085
1505801	CaPO4-	3.100	6.459
1505802	CaH2PO4+	-1.120	20.960
1601300	CdBr+	-0.810	2.170
1601301	CdBr2	0.000	2.899

Table 32.   Aqueous species for which MINTEQ version 3 enthalpy and logK values are identical to those in wateq4f.dat, a database distributed with PHREEQC.—Continued

MINTEQ ID	Species	Enthalpy [kilocalorie per mole]	LogK
1601800	CdCl+	0.590	1.980
1601801	CdCl2	1.240	2.600
1601802	CdCl3-	3.900	2.399
1601803	CdOHCl	4.355	-7.404
1602700	CdF+	0.000	1.100
1602701	CdF2	0.000	1.500
1603300	CdOH+	13.100	-10.080
1603301	Cd(OH)2	0.000	-20.350
1603302	Cd(OH)3-	0.000	-33.300
1603303	Cd(OH)4-2	0.000	-47.350
1603304	Cd2OH+3	10.899	-9.390
1604920	CdNO3+	-5.200	0.399
1607320	CdSO4	1.080	2.460
1607321	Cd(SO4)2-2	0.000	3.500
2301800	CuCl2-	-0.420	5.500
2301801	CuCl3-2	0.260	5.700
2311400	CuCO3	0.000	6.730
2311401	Cu(CO3)2-2	0.000	9.830
2311800	CuCl+	8.650	0.430
2311801	CuCl2	10.560	0.160
2311802	CuCl3-	13.690	-2.290
2312700	CuF+	1.620	1.260
2313300	CuOH+	0.000	-8.000
2313301	Cu(OH)2	0.000	-13.680
2313302	Cu(OH)3-	0.000	-26.899
2313303	Cu(OH)4-2	0.000	-39.600
2313304	Cu2(OH)2+2	17.539	-10.359
2317320	CuSO4	1.220	2.310
2803300	FeOH+	13.199	-9.500

Table 32. Aqueous species for which MINTEQ version 3 enthalpy and logK values are identical to those in wateq4f.dat, a database distributed with PHREEQC.—Continued

MINTEQ ID	Species	Enthalpy [kilocalorie per mole]	LogK
2803301	FeOH3-1	30.300	-31.000
2803302	FeOH2	28.565	-20.570
2807300	Fe(HS)2	0.000	8.950
2807301	Fe(HS)3-	0.000	10.987
2807320	FeSO4	3.230	2.250
2811800	FeCl+2	5.600	1.480
2811801	FeCl2+	0.000	2.130
2811802	FeCl3	0.000	1.130
2812700	FeF+2	2.699	6.199
2812701	FeF2+	4.800	10.800
2812702	FeF3	5.399	14.000
2813300	FeOH+2	10.399	-2.190
2813304	Fe2(OH)2+4	13.500	-2.950
2813305	Fe3(OH)4+5	14.300	-6.300
3300603	H4AsO3+	0.000	-0.305
3300900	H2BO3-1	3.224	-9.240
3302702	H2F2	0.000	6.768
3304900	NH3	12.480	-9.252
3305800	HPO4-2	-3.530	12.346
3305801	H2PO4-	-4.520	19.553
3307300	H2S	-5.300	6.994
3307301	S-2	12.100	-12.918
4107320	KSO4-	2.250	0.850
4407320	LiSO4-	0.000	0.640
4605800	MgPO4-	3.100	6.589
4605801	MgH2PO4+	-1.120	21.066
4605802	MgHPO4	-0.230	15.220
4703300	MnOH+	14.399	-10.590
4703301	Mn(OH)3-1	0.000	-34.800

Table 32. Aqueous species for which MINTEQ version 3 enthalpy and logK values are identical to those in wateq4f.dat, a database distributed with PHREEQC.—Continued

MINTEQ ID	Species	Enthalpy [kilocalorie per mole]	LogK
4704920	Mn(NO3)2	-0.396	0.600
4907320	NH4SO4-	0.000	1.110
5001400	NaCO3-	8.911	1.268
5007320	NaSO4-	1.120	0.700
5401300	NiBr+	0.000	0.500
5401401	NiCO3	0.000	6.870
5401402	Ni(CO3)2-2	0.000	10.110
5401800	NiCl+	0.000	0.399
5401801	NiCl2	0.000	0.960
5402700	NiF+	0.000	1.300
5403300	NiOH+	12.420	-9.860
5403301	Ni(OH)2	0.000	-19.000
5403302	Ni(OH)3-	0.000	-30.000
5407320	NiSO4	1.520	2.290
5407321	Ni(SO4)2-2	0.000	1.020
6001300	PbBr+	2.880	1.770
6001301	PbBr2	0.000	1.440
6001400	Pb(CO3)2-2	0.000	10.640
6001401	PbCO3	0.000	7.240
6001800	PbCl+	4.380	1.600
6001801	PbCl2	1.080	1.800
6001802	PbCl3-	2.170	1.699
6001803	PbCl4-2	3.530	1.380
6002700	PbF+	0.000	1.250
6002701	PbF2	0.000	2.560
6002702	PbF3-	0.000	3.420
6002703	PbF4-2	0.000	3.100
6003300	PbOH+	0.000	-7.710
6003301	Pb(OH)2	0.000	-17.120

Table 32.   Aqueous species for which MINTEQ version 3 enthalpy and logK values are identical to those in wateq4f.dat, a database distributed with PHREEQC.—Continued

MINTEQ ID	Species	Enthalpy [kilocalorie per mole]	LogK
6003302	Pb(OH)3-	0.000	-28.060
6003303	Pb2OH+3	0.000	-6.360
6003304	Pb3(OH)4+2	26.500	-23.880
6003305	Pb(OH)4-2	0.000	-39.699
6004920	PbNO3+	0.000	1.170
6007320	PbSO4	0.000	2.750
6007321	Pb(SO4)2-2	0.000	3.470
7702700	SiF6-2	-16.260	30.180
8913301	U(OH)2+2	17.730	-2.270
8913302	U(OH)3+	22.645	-4.935
8913303	U(OH)4	24.760	-8.498
8935804	UO2H2PO4)3	-28.600	66.245
9501300	ZnBr+	0.000	-0.580
9501301	ZnBr2	0.000	-0.980
9501401	ZnCO3	0.000	5.300
9501402	Zn(CO3)2-2	0.000	9.630
9501800	ZnCl+	7.790	0.430
9501801	ZnCl2	8.500	0.450
9501802	ZnCl3-	9.560	0.500
9501803	ZnCl4-2	10.960	0.199
9501804	ZnOHCl	0.000	-7.480
9502700	ZnF+	2.220	1.150
9503300	ZnOH+	13.399	-8.960
9503301	Zn(OH)2	0.000	-16.899
9503302	Zn(OH)3-	0.000	-28.399
9503303	Zn(OH)4-2	0.000	-41.199
9507320	ZnSO4	1.360	2.370
9507321	Zn(SO4)2-2	0.000	3.280

Table 33. Aqueous species for which MINTEQ version 3 enthalpy and logK values differed from those in wateq4f.dat, a database distributed with PHREEQC. Enthalpy and logK values have been changed in OTEQ's thermo.dbs file to match those in wateq4f.dat.

[kcal/mole, kilocalories per mole]

MINTEQ ID	Species	MINTEQ ver. 3 (Allison and others, 1991)		MINTEQ ver. 4 (HydroGeoLogic, Inc., & Allison Geosci. Consultants, 1999)		OTEQ (based on wateq4f.dat, Parkhurst and Appelo, 1999)	
		Enthalpy [kcal/mole]	LogK	Enthalpy [kcal/mole]	LogK	Enthalpy [kcal/mole]	LogK
302700	AlF+2	0.000	7.010	1.099	7.000	1.060	7.000
302701	AlF2+	20.000	12.750	1.984	12.600	1.980	12.700
302702	AlF3	2.500	17.020	2.079	16.700	2.160	16.800
302703	AlF4-	0.000	19.720	2.079	19.400	2.200	19.400
303300	AlOH+2	11.899	-4.990	11.430	-4.997	11.490	-5.000
303301	Al(OH)2+	0.000	-10.100	0.000	-10.094	26.900	-10.100
303302	Al(OH)4-	44.060	-23.000	41.405	-22.688	42.300	-22.700
303303	Al(OH)3	0.000	-16.000	0.000	-16.791	39.890	-16.900
307320	AlSO4+	2.150	3.020	6.692	3.890	2.290	3.500
307321	Al(SO4)2-	2.840	4.920	2.840	4.920	3.110	5.000
1507320	CaSO4	1.470	2.309	1.697	2.360	1.650	2.300
2311803	CuCl4-2	7.780	-4.590	7.780	-4.590	17.780	-4.590
2813301	FeOH2+	0.000	-5.670	0.000	-4.594	17.100	-5.670
2813302	FeOH3	0.000	-13.600	24.800	-12.560	24.800	-12.560
2813303	FeOH4-	0.000	-21.600	0.000	-21.588	31.900	-21.600
2817320	FeSO4+	3.910	3.920	5.975	4.050	3.910	4.040
2817321	Fe(SO4)2-	4.600	5.420	4.590	5.380	4.600	5.380
3300600	H2AsO3-	6.560	-9.228	6.551	-9.290	6.580	-9.150
3300601	HAsO3-2	14.199	-21.330	14.199	-21.330	14.199	-23.850
3300602	AsO3-3	20.250	-34.744	20.250	-34.744	20.250	-39.550
3300611	H2AsO4-	-1.690	-2.243	-1.697	-2.240	-1.690	-2.300
3300612	HAsO4-2	-0.920	-9.001	-0.980	-9.200	-0.920	-9.460
3300613	AsO4-3	3.430	-20.597	3.083	-20.700	3.430	-21.110
3301400	HCO3-	-3.617	10.330	-3.489	10.329	-3.561	10.329
3301401	H2CO3	-2.247	16.681	-5.679	16.681	-5.738	16.681
3302700	HF	3.460	3.169	3.179	3.170	3.180	3.180
3307620	HSeO4-1	4.200	1.906	5.497	1.700	4.910	1.660
3307700	H3SiO4-	8.935	-9.930	4.780	-9.840	6.120	-9.830
3307701	H2SiO4-2	29.714	-21.619	14.579	-23.040	17.600	-23.000
4603300	MgOH+	15.935	-11.790	16.207	-11.397	15.952	-11.440
4701801	MnCl2	0.000	0.041	0.000	0.250	0.000	0.250
4701802	MnCl3-	0.000	-0.305	0.000	-0.310	0.000	-0.310
5002700	NaF	0.000	-0.790	2.868	-0.200	0.000	-0.240

Table 34. Mineral species for which MINTEQ version 3 enthalpy and logK values are identical to those in wateq4f.dat, a database distributed with PHREEQC.[1]

MINTEQ ID	Species	Enthalpy [kilocalorie per mole]	LogK
16001	GAMMA CD	18.140	-13.590
1002000	ACANTHITE	-53.300	36.050
1016000	GREENOCKITE	-16.360	15.930
1023000	CHALCOCITE	-49.350	34.619
1023001	DJURLEITE	-47.881	33.920
1023002	Anilite	-43.535	31.878
1023003	BLAUBLEI II	0.000	27.279
1023100	BLAUBLEI I	0.000	24.162
1023102	CHALCOPYRITE	-35.480	35.270
1028000	FES PPT	0.000	3.915
1028001	GREIGITE	0.000	45.035
1028002	MACKINAWITE	0.000	4.648
1028003	PYRITE	-11.300	18.479
1047000	MNS GREEN	5.790	-3.800
1054000	MILLERITE	-2.500	8.042
1095000	ZNS (A)	-3.670	9.052
1095001	SPHALERITE	-8.250	11.618
1095002	WURTZITE	-5.060	9.682
2002000	AG2O	10.430	-12.580
2016000	CD(OH)2 (A)	20.770	-13.730
2016001	CD(OH)2 (C)	0.000	-13.650
2023000	CUPRITE	-6.245	1.550
2023100	CU(OH)2	15.250	-8.640
2023101	TENORITE	15.240	-7.620
2023102	DIOPTASE	8.960	-6.500
2028100	FERRIHYDRITE	0.000	-4.891
2028101	FE3(OH)8	0.000	-20.222

Table 34. Mineral species for which MINTEQ version 3 enthalpy and logK values are identical to those in wateq4f.dat, a database distributed with PHREEQC.[1]—Continued

MINTEQ ID	Species	Enthalpy [kilocalorie per mole]	LogK
2054000	NI(OH)2	-30.450	-10.800
2054001	BUNSENITE	23.920	-12.450
2060000	MASSICOT	16.780	-12.910
2060001	LITHARGE	16.380	-12.720
2060002	PBO,.3H2O	0.000	-12.980
2060003	PLATTNERITE	70.730	-49.300
2060004	PB(OH)2 (C)	13.990	-8.150
2060005	PB2O(OH)2	0.000	-26.200
2077001	CRISTOBALITE	-5.500	3.587
2077003	SIO2(A,GL)	-4.440	3.018
2089300	UO3 (C)	19.315	-7.719
2089301	GUMMITE	23.015	-10.403
2089302	B-UO2(OH)2	13.730	-5.544
2089303	SCHOEPITE	12.045	-5.404
2095000	ZN(OH)2 (A)	0.000	-12.450
2095001	ZN(OH)2 (C)	0.000	-12.200
2095002	ZN(OH)2 (B)	0.000	-11.750
2095003	ZN(OH)2 (G)	0.000	-11.710
2095004	ZN(OH)2 (E)	0.000	-11.500
2095005	ZNO(ACTIVE)	0.000	-11.310
2095006	ZINCITE	21.860	-11.140
3006100	AS2O5	5.405	-6.699
3023000	CUPROUSFERIT	3.800	8.920
3023100	CUPRICFERIT	38.690	-5.880
3028000	MAGNETITE	50.460	-3.737
3028100	HEMATITE	30.845	4.008
3028101	MAGHEMITE	0.000	-6.386
3047100	BIXBYITE	15.245	0.611

Table 34. Mineral species for which MINTEQ version 3 enthalpy and logK values are identical to those in wateq4f.dat, a database distributed with PHREEQC.[1]—Continued

MINTEQ ID	Species	Enthalpy [kilocalorie per mole]	LogK
3050000	NATRON	-15.745	1.311
3060000	PB2O3	0.000	-61.040
3060001	MINIUM	102.760	-73.690
3089100	U4O9 (C)	101.235	3.384
4002000	BROMYRITE	-20.170	12.270
4016000	CDBR2,4H2O	-7.230	2.420
4023000	CUBR	-13.080	8.210
4060000	PBBR2	-8.100	5.180
4060001	PBBRF	0.000	8.490
4095000	ZNBR2, 2H2O	7.510	-5.210
4102000	CERARGYRITE	-15.652	9.750
4116000	CDCL2	4.470	0.680
4116001	CDCL2,1H2O	1.820	1.710
4116002	CDCL2,2.5H2O	-1.710	1.940
4116003	CDOHCL	7.407	-3.520
4123000	NANTOKITE	-9.980	6.760
4123100	MELANOTHALLI	12.320	-3.730
4123101	ATACAMITE	18.690	-7.340
4128100	FEOH)2.7CL.3	0.000	3.040
4147000	MNCL2,4H2O	-17.380	-2.710
4150000	HALITE	-0.918	-1.582
4160000	COTUNNITE	-5.600	4.770
4160001	MATLOCKITE	-7.950	9.430
4160002	PHOSGENITE	0.000	19.810
4160003	LAURIONITE	0.000	-0.623
4160004	PB2(OH)3CL	0.000	-8.793
4195000	ZNCL2	17.480	-7.030
4195001	ZN2(OH)3CL	0.000	-15.200

Table 34.   Mineral species for which MINTEQ version 3 enthalpy and logK values are identical to those in wateq4f.dat, a database distributed with PHREEQC.[1]—Continued

MINTEQ ID	Species	Enthalpy [kilocalorie per mole]	LogK
4195002	ZN5(OH)8CL2	0.000	-38.500
4202000	AGF.4H2O	-4.270	-0.550
4210000	BAF2	-1.000	5.760
4216000	CDF2	9.720	2.980
4223000	CUF	12.370	-7.080
4223100	CUF2	13.320	0.620
4223101	CUF2,2H2O	3.650	4.550
4260000	PBF2	0.700	7.440
4280000	SRF2	-1.250	8.540
4289100	UF4 (C)	18.900	18.606
4289101	UF4.2.5H2O	0.588	27.570
4295000	ZNF2	13.080	1.520
4302000	IODYRITE	-26.820	16.070
4306000	ASI3	-1.875	-4.155
4316000	CDI2	-4.080	3.610
4323000	CUI	-20.140	11.890
4360000	PBI2	-15.160	8.070
4395000	ZNI2	13.440	-7.230
5002000	AG2CO3	-9.530	11.070
5015003	HUNTITE	25.760	29.968
5023100	CUCO3	0.000	9.630
5046000	ARTINITE	28.742	-9.600
5046002	MAGNESITE	6.169	8.029
5046003	NESQUEHONITE	5.789	5.621
5050001	THERMONATR	2.802	-0.125
5054000	NICO3	9.940	6.840
5060000	CERRUSITE	-4.860	13.130
5060001	PB2OCO3	11.460	0.500

Table 34. Mineral species for which MINTEQ version 3 enthalpy and logK values are identical to those in wateq4f.dat, a database distributed with PHREEQC.[1]—Continued

MINTEQ ID	Species	Enthalpy [kilocalorie per mole]	LogK
5060002	PB3O2CO3	26.430	-11.020
5060003	HYDCERRUSITE	0.000	17.460
5095000	SMITHSONITE	4.360	10.000
5095001	ZNCO3,1H2O	0.000	10.260
5123100	CU2(OH)3NO3	17.350	-9.240
5216000	CD(BO2)2	0.000	-9.840
5260000	PB(BO2)2	5.800	-7.610
5295000	ZN(BO2)2	0.000	-8.290
6002000	AG2SO4	-4.250	4.920
6016000	CD3(OH)4SO4	0.000	-22.560
6016001	CD3OH2(SO4)2	0.000	-6.710
6016002	CD4(OH)6SO4	0.000	-28.400
6016003	CDSO4	14.740	0.100
6016004	CDSO4, 1H2O	7.520	1.657
6016005	CDSO4,2.7H2O	4.300	1.873
6023000	CU2SO4	4.560	1.950
6023100	ANTLERITE	0.000	-8.290
6023101	BROCHANTITE	0.000	-15.340
6023102	LANGITE	39.610	-16.790
6023103	CUOCUSO4	35.575	-11.530
6023104	CUSO4	18.140	-3.010
6023105	CHALCANTHITE	-1.440	2.640
6041000	ALUM K	-7.220	5.170
6046000	EPSOMITE	-2.820	2.140
6047000	MNSO4	15.480	-2.669
6047100	MN2(SO4)3	39.060	5.711
6050001	MIRABILITE	-18.987	1.114
6050002	THENARDITE	0.572	0.179

Table 34. Mineral species for which MINTEQ version 3 enthalpy and logK values are identical to those in wateq4f.dat, a database distributed with PHREEQC.[1]—Continued

MINTEQ ID	Species	Enthalpy [kilocalorie per mole]	LogK
6054000	NI4(OH)6SO4	0.000	-32.000
6054001	RETGERSITE	-1.100	2.040
6054002	MORENOSITE	-2.940	2.360
6060000	LARNAKITE	6.440	0.280
6060001	PB3O2SO4	20.750	-10.400
6060002	PB4O3SO4	35.070	-22.100
6060003	ANGLESITE	-2.150	7.790
6060004	PB4(OH)6SO4	0.000	-21.100
6095000	ZN2(OH)2SO4	0.000	-7.500
6095001	ZN4(OH)6SO4	0.000	-28.400
6095002	ZN3O(SO4)2	62.000	-19.020
6095003	ZINCOSITE	19.200	-3.010
6095004	ZNSO4, 1H2O	10.640	0.570
6095005	BIANCHITE	0.160	1.765
6095006	GOSLARITE	-3.300	1.960
7002000	AG3PO4	0.000	17.550
7010000	URANOCIRCITE	10.100	44.631
7015000	NINGYOITE	2.270	53.906
7015001	AUTUNITE	14.340	43.927
7015002	FCO3APATITE	-39.390	114.400
7016000	CD3(PO4)2	0.000	32.600
7023100	CU3(PO4)2	0.000	36.850
7023101	CU3(PO4)2,3W	0.000	35.120
7023102	TORBERNITE	15.900	45.279
7028000	BASSETITE	19.900	44.485
7028001	VIVIANITE	0.000	36.000
7028100	STRENGITE	2.030	26.400
7041000	K-AUTUNITE	-5.860	48.244

Table 34. Mineral species for which MINTEQ version 3 enthalpy and logK values are identical to those in wateq4f.dat, a database distributed with PHREEQC.[1]—Continued

MINTEQ ID	Species	Enthalpy [kilocalorie per mole]	LogK
7046000	SALEEITE	20.180	43.646
7047000	MN3(PO4)2	-2.120	23.827
7049000	URAMPHITE	-9.700	51.749
7050000	NA-AUTUNITE	0.460	47.409
7054000	NI3(PO4)2	0.000	31.300
7060000	PRZHEVALSKIT	11.000	44.365
7060001	CLPYROMORPH	0.000	84.430
7060002	HXYPYROMORPH	0.000	62.790
7060003	PLUMBGUMMITE	0.000	32.790
7060004	HINSDALITE	0.000	2.500
7060005	TSUMEBITE	0.000	9.790
7080000	SR-AUTUNITE	13.050	44.457
7089301	H-AUTUNITE	3.600	47.931
7095000	ZN3(PO4),4W	0.000	32.040
8015000	URANOPHANE	0.000	-17.490
8054000	NI2SIO4	33.360	-14.540
8060000	PB2SIO4	26.000	-19.760
8095000	WILLEMITE	33.370	-15.330
8216000	CDSIO3	16.630	-9.060
8260000	PBSIO3	9.260	-7.320
8295000	ZNSIO3	18.270	-2.930
8450000	MAGADIITE	0.000	14.300
8628000	GREENALITE	0.000	-20.810

[1]Actual values in wateq4f.dat are equal to -1 times the values shown in the table, due to the different way the mineral reactions are written within PHREEQC.

Table 35.   Mineral species for which MINTEQ version 3 enthalpy and logK values differ from those in wateq4f.dat, a database distributed with PHREEQC. Default enthalpy and logK values in OTEQ's type6.dbs file are equal to the MINTEQ version 3 values.

[kcal/mole, kilocalories per mole]

MINTEQ ID	Species	PHREEQC (wateq4f.dat, Parkhurst and Appelo, 1999)[1]		OTEQ (MINTEQ ver. 3, Allison and others, 1991)	
		Enthalpy [kcal/mole]	LogK	Enthalpy [kcal/mole]	LogK
2003000	ALOH3(A)	26.500	-10.800	27.045	-10.380
6041001	ALUNITE	50.250	1.400	-3.918	1.346
6015000	ANHYDRITE	1.710	4.360	3.769	4.637
5015000	ARAGONITE	2.589	8.336	2.615	8.360
6010000	BARITE	-6.350	9.970	-6.280	9.976
2003001	BOEHMITE	28.181	-8.584	28.130	-8.578
2046000	BRUCITE	27.100	-16.840	25.840	-16.792
5015001	CALCITE	2.297	8.480	2.585	8.475
6080000	CELESTITE	1.037	6.630	0.470	6.465
2077000	CHALCEDONY	-4.720	3.550	-4.615	3.523
8646000	CHRYSOTILE	46.800	-32.200	52.485	-32.188
8246000	CLINOENSTITE	20.049	-11.342	20.015	-11.338
2003002	DIASPORE	24.681	-6.879	24.630	-6.873
8215000	DIOPSIDE	32.348	-19.894	32.280	-19.886
5015002	DOLOMITE	9.436	17.090	8.290	17.000
4215000	FLUORITE	-4.690	10.600	-4.710	10.960
8046000	FORSTERITE	48.578	-28.306	48.510	-28.298
2003003	GIBBSITE(C)	22.800	-8.110	22.800	-8.770
2028102	GOETHITE	14.480	1.000	14.480	-0.500
6015001	GYPSUM	0.109	4.580	-0.261	4.848
3047000	HAUSMANNITE	100.640	-61.030	80.140	-61.540
5046001	HYDRMAGNESIT	52.244	8.762	52.210	8.766
8603001	KAOLINITE	35.300	-7.435	35.280	-5.726
6028000	MELANTERITE	-4.910	2.209	-2.860	2.470

Table 35.   Mineral species for which MINTEQ version 3 enthalpy and logK values differ from those in wateq4f.dat, a database distributed with PHREEQC. Default enthalpy and logK values in OTEQ's type6.dbs file are equal to the MINTEQ version 3 values.—Continued

[kcal/mole, kilocalories per mole]

MINTEQ ID	Species	PHREEQC (wateq4f.dat, Parkhurst and Appelo, 1999)[1]		OTEQ (MINTEQ ver. 3, Allison and others, 1991)	
		Enthalpy [kcal/mole]	LogK	Enthalpy [kcal/mole]	LogK
2015001	PORTLANDITE	31.000	-22.800	30.690	-22.675
2047003	PYROCROITE		-15.200	22.590	-15.088
2077002	QUARTZ	-5.990	3.980	-6.220	4.006
1006001	REALGAR	-30.894	19.944	-30.545	19.747
5089300	RUTHERFORDIN	1.440	14.450	1.440	14.439
8646003	SEPIOLITE(C	10.700	-15.760	27.268	-15.913
2077004	SIO2(A,PT)	-3.340	2.710	-3.910	2.710
5080000	STRONTIANIT	0.400	9.271	0.690	9.250
8215001	TREMOLITE	96.853	-56.574	96.615	-56.546
3089101	U3O8(C)	116.000	-20.530	116.020	-21.107
2089100	URANINITE	18.610	4.800	18.630	4.700
5010000	WITHERITE	-0.703	8.562	-0.360	8.585

[1]Actual values in wateq4f.dat are equal to −1 times the values shown in the table, due to the different way the mineral reactions are written within PHREEQC.

www.ingramcontent.com/pod-product-compliance
Lightning Source LLC
Chambersburg PA
CBHW081504170526
45166CB00008B/2547